Social Change and Scientific Organization

The Royal Institution, 1799–1844

Frontispiece: The Royal Institution, ca. 1838

Social Change and Scientific Organization

The Royal Institution, 1799–1844

Morris Berman

CORNELL UNIVERSITY PRESS

ITHACA, NEW YORK

First published 1978 by Cornell University Press.

International Standard Book Number 0-8014-1093-2
Library of Congress Catalog Number 77-79702
Printed in the United States of America.

To

Joseph Bitensky
(1878/9–1965)

Contents

List of Plates

Acknowledgements

The Royal Institution: 1c, 2, 9b, 11, 14d, 15a, 17a-b, 18a-b, 19, 20, 21, 24, 25,
    26, 27, 28,
The British Museum: 1b, 3, 6, 7, 9a, 13a, 14b, 16
The British Library Board: 4b, 8
The National Portrait Gallery, London: 1a, 4a, 5a-b, 13b, 14a, c
Bedford Estates: 10
Science Museum, London: 22
Plate No. 12 from Penlee House Museum, is reproduced with the kind permis-
    sion of the Town Mayor and Charter Trustees of Penzance
Plate No. 23 is reproduced by permission of the Master and Fellows of
    Trinity College, Cambridge

Preface

This book is not, despite the subtitle, a 'history' of the Royal Institution in the commonly accepted meaning of the term. It is, rather, an attempt to use England's first major scientific laboratory to ask what I hope are significant questions regarding the nature and function of science in industrial society. Subjects such as finances, building construction, or day-to-day activities receive brief treatment, whereas the social background, and the Institutional activities rooted in that background, occupy the bulk of my analysis. I have not, in short, tried to 'recapture' the Royal Institution of 1799–1844 as much as I have tried to clarify its role in the creation of modern scientific society, and the objective conditions that made that role possible.

It is a truism that historiography tends to reflect contemporary concerns. For the Victorian physician Henry Bence Jones, who wrote the only other full-length study of the Royal Institution, science, the RI, and indeed the whole British empire were 'built to last'. One hundred years later, the eclipse of British hegemony seems a minor issue in the light of speculations on the coming of post-industrial society; nor are we so certain, in the mid-1970s, of the Victorian ethos that progress is inevitable and that science/technology delivers the goods. In the face of an increasing awareness of the abusive nature of science and technology—whether this involves ecological spoliation, overemphasis on behaviour modification and the general monitoring of human life, psychological alienation, or increasing technological chaos—the whole question of the early organization of science ceases to be merely an academic one. During the past few years there has been a growing feeling in the Western industrial nations that something philosophical, or even metaphysical, rather than just economic, is definitely awry with our way of life, and that science and technology may be more a part of the problem than a part of the solution. If this is true, then some understanding of the premises on which institutional science was organized, and of how science came to permeate the value systems of capitalist societies, is clearly essential. A study of one scientific institution cannot, of course, hope to solve such immense issues; but it does seem to me that the Royal Institution was (to use an overworked term) highly paradigmatic in its impact on the first industrial nation. It is the legacy of that impact, which seems (from the vantage point of the late twentieth century) both inevitable and disturbing, that has motivated me to undertake this work.

The empirical data on which the following analysis of the Royal Institution is based are contained in a set of six appendices, which unfortunately had to be omitted from the text. It was my original intention that these be included in the published work, but the length of this material—more than one hundred single-spaced typewritten pages—made it virtually impossible to do so, given the sizeable increase in the cost of production which this would have entailed. This material includes biographical profiles of the hereditary Proprietors of the RI (357 in all), the RI governors (Managers and Visitors) for the period 1799–1840 (235 in all), and the Members elected during 1826–7 (129), as well as statistical analyses of this information and a set of graphs. There is also a discussion of conflicting evidence on Count Rumford, and a statistical profile of the Friday Evening Discourses for 1826–40. By and large the appendices, being for the most part straightforward mechanical compilations, are not inherently interesting, and from this point of view their omission is no great loss. Nevertheless, as the thesis presented below is a controversial one, professional sociologists and historians of science would perhaps have wanted easy access to the empirical data, and I can only regret that this was not possible. I have, instead, deposited copies of this material at the Royal Institution and at the library of The Johns Hopkins University in Baltimore, and can only hope that this will, in part, compensate for the omission. In any event, the reader should be aware that all discussions of the statistical composition of the Proprietorship, Membership, or governorship of the RI (especially in terms of social class or group affiliation) which occur at various points throughout the text are taken from these appendices.

Research on this topic was supported over the years by the Rutgers Research Council, the National Science Foundation, and the Leverhulme Trust. I am also grateful to a number of institutions for permission to cite material from their archives: the Royal Society of London; the Royal Institution of Great Britain; the Royal Agricultural Society of England; the John Rylands University Library of Manchester; the Matthew Boulton Trust; the Library of University College London; and the Institution of Electrical Engineers. Unpublished Crown Copyright material in the India Office Library and Records reproduced in this book appears by permission of the Controller of Her Majesty's Stationery Office. Chapter 2 first appeared, in its essentials, in an article in *Social Studies of Science* (formerly *Science Studies*), Vol. 2, Number 3, July 1972, and is reprinted here by permission of the publishers, Sage Publications (London and Beverly Hills).

My greatest debt is to Robert H. Kargon of The Johns Hopkins University, who first suggested to me the importance of institutional history, and the methodology by which such so-called 'externalist' studies could best be pursued. I think it fair to say that his own feel for crucial issues and scholarly explication was the most formative influence on my intellectual career. Professor Kargon was kind enough to read the final version of this work and offer insightful criticism and suggestions; responsibility for errors of fact or interpretation, however, is strictly my own. I am also very grateful to Peter N.

Stearns of Carnegie-Mellon University for his comments on this work. Much of my thinking over the years has been influenced by discussions with Steven J. Cagney and David Kubrin, and although their help was of a general nature rather than directed to this particular manuscript, I doubt whether my conceptualization of the issues raised in this work would have been the same without their friendship. I was fortunate to have the invaluable assistance of Jeremy Weston and James Friday at the Royal Institution, and of Lenore Simons at the Institution of Electrical Engineers. W. H. Austin and Sophie Forgan assisted me at various times with the identification of the Royal Institution's governors. Finally, Lois Haggerty provided a great deal of emotional support during the last two years this work was being written, and it is a joy to have her constant companionship.

San Francisco   M.B.

January 1976

Abbreviations Used in the Text

AY	Arthur Young (ed.), *Annals of Agriculture, and other Useful Arts*, 46 vols., 1784–1815
BJ	Henry Bence Jones, *The Royal Institution: Its Founder and Its First Professors* (London: Longmans, Green, 1871)
BJ, I, II	Henry Bence Jones, *The Life and Letters of Faraday* (2 vols.; London: Longmans, Green, 1870). There were two editions in 1870; all citations refer to the first.
BM Add. MSS.	British Museum, Additional Manuscripts
DNB	*Dictionary of National Biography*
DW	John Davy (ed.), *The Collected Works of Sir Humphry Davy, Bart.* (9 vols.; London: Smith, Elder, 1839)
EIC	East India Company
IEE	Institution of Electrical Engineers
IOL	India Office Library and Records
JAP, I, II	John A. Paris, *The Life of Sir Humphry Davy, Bart.* (2 vols.; London: Henry Colburn and Richard Bentley, 1831)
JD, I, II	John Davy, *Memoirs of the Life of Sir Humphry Davy, Bart.* (2 vols.; London: Longman, Rees, Orme, Brown, Green, and Longman, 1836)
LI	London Institution
LPW	L. Pearce Williams, *Michael Faraday* (New York: Basic Books, 1965)
LPW, I, II	L. Pearce Williams (ed.), *The Selected Correspondence of Michael Faraday* (2 vols.; Cambridge: University Press, 1971)
MM	Minutes of the Meetings of the Managers of the Royal Institution, MS volumes at the Royal Institution
Phil. Mag.	*The Philosophical Magazine and Journal*
Phil. Trans.	*The Philosophical Transactions of the Royal Society of London*
QJS	Also referred to as 'Brande's journal' in the text: *The Journal of Science and the Arts*, vols. 1–6 (1816–19) *The Quarterly Journal of Science, Literature, and the Arts,*

	vols. 7–22 (1819–27)
	The Quarterly Journal of Science, Literature, and Art, n.s., vols. 1–7 (1827–30)
	Journal of the Royal Institution of Great Britain, vols. 1–2 (2nd vol. incomplete), 1830–1
RASE	Royal Agricultural Society of England
RI	Royal Institution of Great Britain
RS	Royal Society of London
SBCP *Reports*	Thomas Bernard (ed.), *The Reports of the Society for Bettering the Condition and Increasing the Comforts of the Poor* (6 vols.; London: W. Bulmer, 1798–1815)
VM	Minutes of the Committee of Visitors of the Royal Institution, MS volumes at the Royal Institution, Items 117 and 118

Introduction

Several years ago Theodore Roszak published a broadside against Western science entitled *The Making of a Counter Culture*. Imbued with a Blakeian vision, the argument was often vague and emotional, and probably made headway only with a counter culture that was already persuaded by the romance in advance. And yet, like so many works of the sixties, it contained a notion that was annoyingly refractory, the pea under the mattresses that could nevertheless still be felt: science was a mythology like any other, and as such, its days were numbered. By a mythology, Roszak meant 'an arbitrary construct in which a given society in a given historical situation has invested its sense of meaningfulness and value'.[1]

The present book is about science as an ideology rather than a mythology, the latter issue being far too cosmic for the scope intended here.[2] Yet a few words must be said about Roszak's thesis before we can proceed to analyse the institutionalization of science during the Industrial Revolution. By 'arbitrary' Roszak meant theoretically replaceable, not historically accidental. That is to say, for reasons still unclear, there were very concrete forces at work that turned the European intellectual mainstream toward the Cartesian-Newtonian *gravitation* epistemology, a problem that plagued Max Weber all his life. Weber's thesis that science, Puritanism, and capitalism arose simultaneously during the sixteenth century and rode like a troika from that point on is generally accepted; but the disquieting corollary, that this epistemology, being rooted in a particular historical configuration, is a metaphysics rather than the unique path to truth, has gained very little acceptance indeed. As Julien Freund maintains in his study of Weber.

> [o]nly in the West has science developed in the sense of a body of knowledge possessing universal validity. Elsewhere we find observations of great subtlety, as

---

[1] Theodore Roszak, *The Making of a Counter Culture* (Garden City: Doubleday, 1969), p. 215.

[2] Perhaps the best elaboration of mythology in the social-functional (as opposed to strictly anthropological) sense is Roland Barthes, *Mythologies*, trans. Annette Lavers (New York: Hill and Wang, 1972). I shall discuss ideology at greater length below, but a shorthand distinction between it and mythology might correspond to the distinction between 'problematic', as the term has been employed by Louis Althusser, and the more grandiose term *Weltanschauung* (*see For Marx*, trans. Ben Brewster [New York: Vintage Books, 1970], pp. 253–4).

well as empirical knowledge and profound reflections on life and the universe, and even philosophical and theological insight, but nowhere else do we find rational demonstration on the basis of mathematics or precise experiment conducted in laboratories equipped with accurate instruments of measurement.[3]

To claim epistemological superiority for the West is, from the viewpoint of the sociology of knowledge, philosophically naive; but to concede that Western science is a thought system appropriate to a certain set of historical circumstances is to cast serious doubt on its unique claim to objectivity. Weber, as is well known, chose the West, but never resolved the problem to his own satisfaction.

In recent times, the only extended examination of this problem is the work of the Frankfurt School for Social Research, whose most familiar representative in the U.S. is the controversial philosopher, Herbert Marcuse. In many ways, *One-Dimensional Man* is a study and critique of that which Weber termed *zweckrational*: that which is *purposively* rational, or *technically* rational, but not necessarily reasonable. This quality Marcuse saw as the defining characteristic of Western scientific civilization; hence the emptiness of our lives. As Oskar Kokoschka put it, in the twentieth century reason was finally reduced to mere function, and the result has been a 'metaphysic of mechanization'.[4] The eventual outcome of the Cartesian–Newtonian method, says Marcuse, is a 'technical reason' that exalts expediency as a value at the very time that it claims to be value-free. The legacy is, as Weber pointed out, a rationalized and reified life, run by bureaucracies that have no purpose other than maximizing their own efficiency. This 'logic of domination' is not merely a problem of the abuse of science or its misapplication; it is rather embedded in the methodology itself. By this method, means and ends have become completely scrambled. 'It is my purpose to demonstrate,' writes Marcuse, 'the *internal* instrumentalist character of this scientific rationality by virtue of which it is *a priori* technology, and the *a priori* of a specific technology—namely, technology as a form of social control and domination'.[5] Weber saw no way out: *die Entzauberung der Welt* was inevitable as science advanced. That there was no bridge between science and faith was a true but tragic condition of mankind.[6]

The elaboration of these broader themes still awaits systematic research and investigation; and although such metaphysical issues clearly hover about the present study, they cannot be its special focus. What interests me in this particular work is the historical process by which science became organized so as to serve, almost literally, as the grammar of industrial society: its set of rules,

[3] Julien Freund, *The Sociology of Max Weber*, trans. Mary Ilford (Harmondsworth: Penguin Books, 1972), p. 142.

[4] Oskar Kokoschka, *My Life*, trans. David Britt (London: Thames and Hudson, 1974), p. 198. A general introduction to the work of the Frankfurt School can be found in Martin Jay, *The Dialectical Imagination* (Boston: Little, Brown, 1973).

[5] Herbert Marcuse, *One-Dimensional Man: Studies in the Ideology of Advanced Industrial Society* (boston: Beacon Press, 1966), pp. 157–8.

[6] George Lichtheim, *From Marx to Hegel* (New York: The Seabury Press, 1971), p. 202.

its ideology. As two historians of science recently put it, the 'validity of scientific knowledge by itself hardly accounts for its very rapid assumption of primacy of place in the [W]estern cultural hierarchy'.[7] How, then, *did* science acquire such prominence in Western life? To probe such an issue it will be best to accept the validity of the scientific method, at least for the purpose of discussion, and leave the foregoing epistemological critique as a caveat, that the problem is probably much deeper than the mere sociological analysis attempted in this book can fathom.

'Mere sociological analysis', however, is hardly without its value. Even if we accept Marcuse's argument about the epistemology of science, that it 'projects a world', there is still the problem of how this came to be translated into a social reality. On this latter issue Roszak is again worth quoting. 'The general public', he writes,

> has had to content itself with accepting the decision of experts that what the scientists say is true, that what the technicians design is beneficial. All that remained to be done to turn such an authoritative professionalism into a new regime of bad magicians was for ruling political and economic elites to begin buying up the experts and using them for their own purposes. It is in this fashion that the technocracy has been consolidated.[8]

Although stated somewhat crudely here, this linking of science to expertise and dominant classes was a central theme of Robert Wiebe's very sophisticated description of the rise of a professional technostructure in the U.S.[9] The creation of a new industrial society, says Wiebe, spawned a new middle class whose loyalty was to a professional network. For this group, skill, function, and organization were the determinants of consciousness and identity. At the heart of this process lay a new conception of science:

> 'Science', the basic word that every school of thought claimed and worshipped, also altered in meaning to accommodate the revolution [in values] ... Bureaucratic thought ... made 'science' practically synonymous with 'scientific method'. Science had become a procedure, or an orientation, rather than a body of results.[10]

All the professions, Wiebe continues, began to view their work as 'scientific', and this altered meaning of the term was extended to all areas of human behaviour. What were thorny political issues were now seen as 'technical difficulties', matters for social engineering.[11] 'Science' thus conceived became

[7] S. A. Shapin and Arnold Thackray, 'Prosopography as a Research Tool in History of Science: the British Scientific Community 1700–1900', *History of Science*, XII (1974), 5.

[8] Roszak, *Counter Culture*, p. 263.

[9] R. H. Wiebe, *The Search for Order 1877–1920* (New York: Hill and Wang, 1968).

[10] ibid., p. 147.

[11] This has been so often repeated that it is now regarded as a truism rather than a position held by professionals designed to benefit from widespread belief in it. Even as eminent an historian as Edmundo O'Gorman could write, as recently as 1972, that 'ungovernable cities, pollution, and ecological unbalance [sic] ... are basically technical problems to be technically solved ...'. ('History, Technology and the Pursuit of Happiness', in *Technology and the Frontiers of Knowledge* [Garden City: Doubleday, 1973], p. 95.)

the value system (ideology) of the professions and eventually of American society at large. Translated into politics it meant the end of politics: a scientifically functioning bureaucracy, government by 'science', not by men.

To what extent this description—or Roszak's—applies to the first industrial nation will be a primary concern of this study. I am not (to reiterate) concerned here with the problems of the scientific method itself, or with the origins of this epistemology, but with the concrete social processes that turned this epistemology into the ideology of advanced industrial society. If we take it as a given, then, that Britain and Western Europe, far in advance of industrial 'take-off', had in its possession a mode of perceiving the world in terms of quantifiable control, it is nevertheless clear that this mode of perception did not become the common denominator of Western life until the nineteenth century. If it is commonly assumed that science is the key to material comfort, that technical expertise can effectively deal with social and economic problems, and that, more generally, reality consists of only that which can be quantified, then it is vital to understand the process by which science became a working metaphysics. If such a major issue cannot be disposed of in a single volume—and as the subtitle of this book indicates, my inquiry revolves around a single scientific institution during roughly the first five decades of its existence—some justification must be made for the choice involved as being relevant to the larger issue at hand.

When one thinks, historically, of the organization of science in England, the Royal Society of London logically comes to mind. Its original inspiration, as many have argued, was Baconian, and many of its early interests revolved around technological problems related to mining, navigation, and war.[12] Yet this interest in applied science and technology was very brief, and by 1700 almost non-existent.[13] The Society became the institutional embodiment of what I have described elsewhere as the 'gentleman amateur tradition'.[14] The whole notion of scientific research was closely bound up with the cultural ideal of the English aristocracy, and in fact was part of its hegemonic apparatus. As Erving Goffman has noted, in England 'the ideology of influential classes focused on the amateur player';[15] and extended to scientific research this meant that the latter was pursued as a kind of hobby, a leisure class and leisure time activity. The dominant picture of the Royal Society, according to Merton, 'is not that of a group of "economic men" jointly or severally seeking to improve their economic standing, but one of a band of curious students cooperatively

---

[12] The standard history is Henry Lyons, *The Royal Society 1660–1940* (Cambridge University Press, 1944), and the most up-to-date is Margery Purver, *The Royal Society: Concept and Creation* (London: Routledge and Kegan Paul, 1967). Technological interests of the early years are discussed in Chs. 7–9 of Robert K. Merton, *Science, Technology and Society in Seventeenth-Century England* (New York: Harper and Row, 1970 reprint).

[13] Margaret 'Espinasse, 'The Decline and Fall of Restoration Science', *Past and Present*, No. 14 (1958), 71–89, and Merton's statistics on pp. 46–7 and 204–6.

[14] Morris Berman, ' "Hegemony" and the Amateur Tradition in British Science', *Journal of Social History*, VIII (Winter, 1975), 30–50.

[15] Erving Goffman, 'Role Distance', in *Encounters* (Harmondsworth: Penguin Books, 1972), p. 124.

delving into the arcana of nature'.[16] By 1729, only two years after Newton's death, the RS nearly merged with the Antiquarian Society, from which, in terms of activity and membership, it was almost indistinguishable.[17] Locke's description of natural philosophy in 1690 as something a gentleman must 'look into . . . to fit himself for conversation' remained true for most of the eighteenth century.[18] It was this view of science as diversion that led Charles Babbage to remark, as late as 1851, that the English language still contained no term by which the occupation of a man of science could be expressed.[19] The close association between the English aristocracy and this view of scientific research as a polite avocation was a powerful legacy for English science, and probably the greatest single constraint on scientific professionalization. The gentleman amateur tradition in British science was, as Merton would say, highly 'congruent' to the world of the landed aristocrat, and made alternative conceptions of science seem weak by comparison.

It would be reasonable, then, to regard any major break in this conception of science as a significant development in its history. As will be argued below, this is precisely what makes the evolution of the RI such a watershed: it was organized by a select group of the upper class with a very different ideology of science in mind, and in the course of its history became a vehicle for an alternative vision, by means of which the dissemination and production of scientific research eventually came to be defined. Even in terms of activities alone, it constituted a significant break, for the Royal Society did no scientific research and offered no scientific instruction. Prior to the founding of the Royal Institution research was a matter of arbitrary interest and private initiative, and teaching an activity reserved for private homes or local societies and study groups. The impact of the Institution was to articulate and consolidate ideologies of science which would ultimately usurp the aristocratic one. Certainly the RI was fashionable and elegant,[20] but fashion and elegance were not its goals. By bending science to entrepreneurial and professional purposes, the RI was the opening wedge in a major ideological shift. This is not to say that the Royal Institution accomplished such a change in and of itself, but that it was, in the first half of the nineteenth century, the focus and vehicle for much of this transition.

[16] Merton, *Science, Technology and Society*, p. 166.
[17] Joan Evans, *A History of the Society of Antiquaries* (Oxford University Press, 1956), p. 83.
[18] Quoted in 'Espinasse, 'Decline and Fall of Restoration Science', p. 76.
[19] Charles Babbage, *The Exposition of 1851* (London: John Murray, 1851), p. 189.
[20] This common view of the RI is reproduced in literally everything written about the Institution, including biographies of Davy, Rumford, and Thomas Young, and it would be impractical to give an extended bibliography at this point. The standard work, and only full-length study, is Henry Bence Jones, *The Royal Institution* (1871). Similar views of the RI may be found in G. A. Foote, 'Sir Humphry Davy and his Audience at the Royal Institution', *Isis*, XLIII (1952), 6–12; K. D. C. Vernon, 'The Foundation and Early Years of the Royal Institution', *Proceedings of the Royal Institution of Great Britain*, XXXIX (1963), 364–402; and the following pieces by Thomas Martin: *The Royal Institution* (3d ed., rev.; London: the Royal Institution, 1961); 'Origins of the Royal Institution', *British Journal for the History of Science*, I (1962), 49–63; and 'Early Years at the Royal Institution', *British Journal for the History of Science*, II (1964), 99–115.

More must be said about the term 'ideology of science', since I have used it freely without precise definition, and since it will figure as an important methodological construct in understanding the dynamics of shifting value systems. The essential approach of this work is that of the sociology of knowledge, inasmuch as it constitutes (to quote Karl Mannheim) an attempt 'to observe how and in what form intellectual life at a given historical moment is related to the existing social and political forces'.[21] Intellectual life, in short, is not a given; rather, it is mediated by elites, vested interests, and so on. An intellectual position, or idea system, can be termed an ideology when it acquires an overt political dimension; when it becomes a rationalization of the world and a plan of action. Thus, for Mannheim, an ideology is a complex of ideas which directs the activity of men toward the maintenance of their position in an existing social order.[22] By 'ideology of science', then, I shall mean that part of a complex of ideas which deals with a certain class' or group's attitude toward science, especially as a factor in maintaining the social order or promoting its own interests, economic or otherwise. Wiebe's discussion of an altered shift in the meaning of science concomitant with the rise of a new professional class would be an obvious example, and a similar change took place at the RI in the 1820s; but it should be clear that the gentleman amateur tradition was itself class-based and hardly neutral. It was, as already noted, part of the aristocracy's hegemonic apparatus, what E. P. Thompson has called the 'theater' of the upper class.[23] Since, as the Italian sociologist Antonio Gramsci argued again and again, it is through such symbols that the values of a ruling class permeate society as a whole, the serious disruption of an ideology of science is a signal that massive structural changes are underway.[24] It is for this reason that I see the Royal Institution as fundamental, for as England's first major scientific laboratory it became the locus of a crucial reorientation in the public conception of science. First, the RI, unlike the RS, was comprised at least partly (in Merton's words) of 'a group of "economic men" jointly or severally seeking to improve their economic standing'.[25] The use of science for personal gain in an environment that hitherto regarded it as a type of finishing school education was no minor development in English history. Secondly, and to my mind even more significantly, the RI became an illustration of how the leadership of the nation might be transferred to another social class, and how a redefinition of science is inseparable from such a transition. For ideology

[21] Karl Mannheim, *Ideology and Utopia*, trans. Louis Wirth and Edward Shils (New York: Harcourt, Brace and World, reprint of 1936 ed.), pp. 67n–68n.
[22] ibid., p. xxi.
[23] Paper given at the Anglo-American History Conference, University of London, June 1972. *See also* R. J. White, *Waterloo to Peterloo* (Harmondsworth: Penguin Books, 1957), pp. 51–2.
[24] On Gramsci's model applied to British science *see* above, footnote 14. Gwyn Williams defined 'hegemony' as 'an order in which a certain way of life and thought is dominant, and in which one concept of reality is diffused throughout society in all its institutional manifestations, informing with its spirit all taste, morality, customs, etc.' ('The Concept of "Egemonia" in the Thought of Antonio Gramsci: Some Notes on Interpretation', *The Journal of the History of Ideas*, XXI [1960], 587).
[25] *See* above, p. *xx*.

should never be so narrowly construed as to mean simply 'a belief system rooted in vested interests'. As Frankl, Erikson, and other psychologists have argued, few of us can manage to make our way in the world without a scheme of patterned activity that enables us to make sense of our lives.[26] This latter tendency Clifford Geertz calls the 'strain theory' of ideology, as opposed to the more simplistic interest theory.[27] According to the strain theory, the goal of ideology is to correct social/psychological imbalance, i.e. to flee anxiety. Nor does this form of ideology exist independently of the pursuit of vested interests: one can seek profit/power and equilibrium simultaneously, or even pursue one by means of the other. This switching and interpenetration of ideologies was in fact common at the Royal Institution. The RI was founded midst the fear of revolution, and scientific activity was seen as a means of containing potential disruption. It soon became an agricultural laboratory, science now being given an entrepreneurial slant and also serving to foster a sense of the avant-garde (and thus collective identity) among a certain sector of the aristocracy. By the mid-1820s, its leadership was largely reformist Whig and Utilitarian, and science was both the rationale for expertise (and thus positions of leadership and control) and the motto of a would-be smoothly functioning social order.

There are, then, two ideological shifts to deal with, though they should not be thought of as chronological stages. The amateur tradition managed to survive into the twentieth century; science-as-technology (not to be confused with technology per se) waxed and waned; and the Utilitarian programme of a scientifically administered society (discussed at length in Chapter 4) suffered severe setbacks from the late 1840s. The Royal Institution did not simplistically shift the nation's scientific gears once and for all. What it did do, in its use of various ideologies, was to open the possibility of these as undermining the older tradition, and demonstrate in specific terms how this might be done.

How are ideological shifts best studied on the micro, or institutional, level? The most thoroughgoing method still remains Sir Lewis Namier's technique of collective biography, or 'prosopography': 'the investigation of the common background characteristics of a group of actors in history by means of a collective study of their lives'.[28] Although my argument cannot be automatically derived from the data themselves, a statistical profile of the men who supported the Institution—called 'Proprietors' until 1810—and the men who governed it over the years—'Managers' and 'Visitors'—is in itself remarkably revealing. It enables us to examine to what extent structure influenced function, as well as to chart the evolution of the Institution during one of the most formative periods

[26] Cf. Erikson's definition of ideology on p. 22 of *Young Man Luther* (New York: W. W. Norton, 1962): 'an unconscious tendency underlying religious and scientific as well as political thought: the tendency at a given time to make facts amenable to ideas, and ideas to facts, in order to create a world image convincing enough to support the collective and the individual sense of identity.'

[27] Clifford Geertz, 'Ideology as a Cultural System', in David E. Apter (ed.), *Ideology and Discontent* (New York: The Free Press, 1964), pp. 47–76.

[28] Lawrence Stone, 'Prosopography', *Daedalus*, C (Winter, 1971), 46. It is a tool, Stone continues, that can be used 'to help explain ideological or cultural change . . .'. *See also* the article by Shapin and Thackray cited in footnote 7, above.

of modern history, the transition from an agrarian to an industrial economy. If, for example, we run such an analysis on the fifty-seven men who first convened on 7 March 1799 to form the Royal Institution, nearly 50% fall into the category of 'improving landlords', i.e. members of the peerage and wealthy gentry who were in the forefront of agricultural development and estate exploitation. Nearly 75% of the first governors were in this group, and this remained the case for the rest of the first decade even though by 1803 the landed interest comprised only about 25% of the Proprietorship. Both the style and substance of the Institution reflected these data. Gilt-edged stationery and expensive coffee could be found in the Reading Room; soil samples and fertilizer in the Laboratory; and a constant stream of enthusiastic propaganda for the entrepreneurial ideology of science in the Lecture Theatre.[29]

Turning to the closing years of the 1830s, it hardly seems that we are talking about the same institution. Agricultural improvers comprise less than 6% of the governorship, whereas the professional classes comprise nearly 60%, the bulk of this being from the legal profession. The work of the Institution has a distinctly administrative flavour, both in the streamlining of its internal operations and in a series of projects—gas illumination, medical lectures, water analyses, Parliamentary testimonies—which were part of the Utilitarian push toward modernization and social coordination. This does not mean that the relationship between structure and function was a simple reflex response; but it does suggest that the evolution of organized science, and the common perception of 'what science is all about', cannot be understood without constant reference to changing class structure and vested interests, i.e. to changes external to science itself.

Finally, the issue of 'externality' requires at least passing reference. Although I do believe that in general terms the structure of scientific thought is derivative from the larger culture of which it is a part, I am, as is any other historian of science who shares this view, unclear as to how this could be demonstrated.[30] The interesting exceptions that do exist—studies by Merton, Zilsel, or Foucault—deal exclusively with the early modern period, and (regrettably) have not really generated a discipleship anyway. In the case of the nineteenth century, there is no such literature at all because, among other things, it is probably still too early to grasp modern industrial society 'from the outside', as it were. Consequently, although I have, in this work, offered some guesses regarding Davy and Faraday, there is no obvious relationship between electrical theories of affinity and the Industrial Revolution, and the internal history of chemistry is undoubtedly more helpful on this point. What I *am*

---

[29] I have retained, throughout the text, the RI's custom of capitalizing words that refer to specific parts of the building or classes of personnel; hence Reading Room, Laboratory, Library, Lecture Theatre, Member, Proprietor, Annual Subscriber, Life Subscriber, Manager, Visitor. The term 'governor', which I have coined to refer to either a Manager or a Visitor, will appear in lower case type.

[30] Stephen Toulmin does, however, offer some important methodological comments in his introduction to Allan Janik and Stephen Toulmin, *Wittgenstein's Vienna* (New York: Simon and Schuster, 1973), on having a *Zeitgeist* analysis without the *Geist*, which might be a possible model.

attempting to demonstrate, however, is the process by which technical reason became (as Michel Foucault would say) the prose of our world. What Davy stood for, for example, ultimately made a greater difference for the history of science than what he accomplished with the voltaic pile, for he paved the way for a society in which the 'social construction of reality' proceeded along scientific lines.[31] Despite all the biographies that have been published since his death in 1829, only J. H. Plumb's *England in the Eighteenth Century* comes close to capturing the essential Davy that the historians of ideas have missed:

> ... Davy at the Royal Institution was something of a portent, the clear practical demonstration of the value of 'useful knowledge' to an industrial society .... It is hard now to recapture the intensity of belief and faith that the accumulation of knowledge and practical scientific understanding would bring both to the nation and to the individual wealth, success, and happiness .... In these years science was woven into the fabric of the nation's life.[32]

How this weaving took place, its impact upon the nation and on science itself, are questions that have been unnecessarily ignored; and the following study of the first few decades of the Royal Institution raises these larger issues explicitly.

[31] Peter L. Berger and Thomas Luckmann, *The Social Construction of Reality* (Garden City: Doubleday, 1966).

[32] J. H. Plumb, *England in the Eighteenth Century* (Harmondsworth: Penguin Books, 1950), p. 170. J. G. Crowther's sketch of Davy in *British Scientists of the Nineteenth Century* (London: Routledge and Kegan Paul, 1935) also adopts this point of view.

Social Change and Scientific Organization

The Royal Institution, 1799–1844

1
Foundation of the Royal Institution

In the last decade of the eighteenth century England was in the grip of what Eric Hobsbawm has called the 'dual revolution', i.e. Industrial and French. The former was generating changes which no one, as yet, could understand; the latter produced a general fear of any social change at all. The former increased the gap between rich and poor, while the latter served as a grim reminder of what such a widening chasm could bring. Legislation to accommodate the former involved the breaking of age-old restrictions on commerce and manufacture; legislation to deal with the latter involved the imposing of restrictions on the rights of speech, assembly, and other civil liberties. From the fall of the Bastille to the erection of Crystal Palace, England walked a tightrope between economic expansion and social distress.

For a small segment of English society, however, the dual revolution presented possibilities that tended to coalesce rather than conflict. The French aristocracy was unaccountably rigid, viewing all opportunities for economic development as a contradiction of its own ideals. Its reaction to a rising industrial class was to use 'the very idea of the gentleman as a weapon against the pretensions of the bourgeoisie'.[1] This attitude was common among the English aristocracy as well, but a number of its members proved to be quite adaptable to the new mode of production. This adaptation served to ease the tensions between the aristocracy[2] and the commercial and industrial classes, and it also seemed to offer, in the sphere of agricultural production, a possible solution to the problem of the poorer classes. For production meant food; and if the land could be made to yield more by enclosure and improved husbandry, the problem of agrarian poverty was ultimately tractable. For this group of 'agricultural improvers', or 'improving landlords', the dual revolution was more complementary than contradictory. The application of the entrepreneurial spirit of the Industrial Revolution to agriculture would, they believed, eventually mitigate the sting of rural poverty and thus the threat of social cataclysm.

It should be emphasized that this group of improving landlords was small.

---

[1] David Landes, *The Unbound Prometheus* (Cambridge University Press, 1969), p. 70. *See also* E. J. Hobsbawm, *Industry and Empire* (London: Weidenfeld and Nicolson, 1968), p. 18.

[2] Although this term commonly refers to the peerage, I shall broaden the definition to include all wealthy landowners, untitled as well as titled.

We shall say more about their contribution to agriculture in Chapter 2, but at present it is necessary to point out that progressive farming and rural philanthropy were not serious concerns of most of the upper class. The two organizations dedicated to these matters—the Board of Agriculture and the Society for Bettering the Condition of the Poor—ultimately failed to interest the rest of the aristocracy in innovative husbandry or allotment of land to the poor. The Royal Institution was, in fact, the only organization created by this group which did have a lasting impact. Begun as an attempt to make rural philanthropy 'scientific', the RI quickly and naturally became an institution wherein science was to be directed primarily toward agricultural improvement.

It is not customary to see the RI, the SBCP, and the Board of Agriculture as a triad, but it was the same set of social and economic developments that brought them into being and gave them a similar, if not common agenda; and it was roughly the same group of men who sat on their governing boards. Since the creation of the RI was part of a more general movement, it will be helpful to elaborate on this connection between agricultural improvement and rural poverty before we can discuss the foundation of the RI itself.

That enclosure and improvement would be viewed, in the 1790s, as solutions to the problem of rural poverty seems at first rather curious. After all, the rapid acceleration of enclosure by Acts of Parliament went back to mid-century, and a host of improvements—in implements, drainage, fertilizers, and livestock—dated from this time as well.[3] Yet this was precisely the same period

[3] During the reign of George III (1760–1820) 75% of all the enclosure acts ever passed by Parliament were proposed, enacted, and then effected. On this *see* Rowland E. Prothero, *English Farming Past and Present* (6th ed.; London: Frank Cass, 1961), p. 257, and Peter Mathias, *The First Industrial Nation* (London: Methuen, 1969), p. 73. On agricultural improvement *see* G. E. Mingay, 'Dr Kerridge's "Agricultural Revolution": A Comment', *Agricultural History*, XLIII (1969), 479.

In general, I am following the revisionist argument regarding the 'agricultural revolution', which sees this period as a short phase of agitated activity rather than a genuinely profound transformation of English agriculture, and the improving landlord as atypical of his class. Important elaborations of this viewpoint include E. L. Jones (ed.), *Agriculture and Economic Growth in England 1650–1815* (London: Methuen, 1967); G. E. Mingay, 'The "Agricultural Revolution" in English History: A Reconsideration', *Agricultural History*, XXXVII (1963), 123–33, and his *English Landed Society in the Eighteenth Century* (London: Routledge and Kegan Paul, 1963); J. D. Chambers and G. E. Mingay, *The Agricultural Revolution 1750–1880* (London: B. T. Batsford, 1966). *See also* H. J. Habakkuk, 'Economic Functions of English Landowners in the Seventeenth and Eighteenth Centuries', *Explorations in Entrepreneural History*, VI (1953), 92–102.

I am, however, dubious about the argument as it concerns the social effects of enclosure. For the traditional interpretation of the 'agricultural revolution' *see* Prothero and also his *Pioneers and Progress of English Farming* (London: Longmans, Green, 1888). The classic statement of the orthodox position is J. L. and Barbara Hammond, *The Village Labourer* (4th ed.; London: Longmans, Green, 1966 reprint), and Paul Mantoux's very fine study, *The Industrial Revolution in the Eighteenth Century*, trans. Marjorie Vernon (rev. ed.; New York: Harper and Row, 1961; original French ed. publ. 1928). A critique of the revisionist position on the impact of enclosure was provided by E. P. Thompson in his review of Chambers and Mingay in 'Land of Our Fathers', *Times Literary Supplement*, 16 February 1967, pp. 117–18. *See also* the very important methodological discussion by Barrington Moore in his *Social Origins of Dictatorship and Democracy* (Harmondsworth: Penguin Books, 1966), esp. pp. 22n and 509–23, and by William Lazonick in 'Karl Marx and Enclosures in England', *The Review of Radical Political Economics*, VI (Summer, 1974), 1–59.

which saw a sharp increase in poverty. The Poor Rate, which amounted to £1,531,732 in 1776, had leaped to £5,348,204 by 1802–3.[4] Clearly, or so it would seem, the remedy could only exacerbate the disease. Yet the assumption, especially in the 1790s, was just the reverse. If Malthus was right, and population was outstripping food supply, a more efficient system of farming made possible by the consolidation of lands might be the answer. Although he would later see enclosure as a cause of poverty, Arthur Young (and the rest of the Board of Agriculture) was enthusiastic about it in the 1790s in the belief that it must lead to greater crop yields.[5] The assumption here, that an increase in production is the cure for poverty, is common enough even today, overlooking as it does the issue of how the resulting wealth will be distributed; but for obvious ideological reasons it is a notion with strong appeal to those most likely to benefit from such an increase. Anything that impeded enclosure and improvement was thus anathema to members of the Board, who spent much time trying to cheapen the enclosure process and interest other landowners in agricultural investment and experimentation.

One glaring obstacle to improvement was, however, poverty itself. The escalating Poor Rate made expensive innovation increasingly difficult, as a number of writers pointed out. 'When the poor's rate amounts to ten shillings,' cried Joseph Townsend,

> or even to four shillings in the pound, who will be at the expense of cleaning, fencing, breaking up, manuring, cropping, the waste and barren parts of an estate? Certainly no gentleman can do it with a view to profit . . . . Were it not for this incumbrance agriculture would certainly be pushed much farther than it has ever been, and many thousand acres of the poorer commons, heaths, and moors, would be inclosed and cultivated . . . . A wise politician will study to remove every obstacle which can retard the progress of improvement: but such is the system of our laws, that the greater the distress among the poor, the less will be the inducement to cultivate our more stubborn and unprofitable lands.[6]

We should not be surprised, then, to see every contemporary agricultural society concerned with the problem of the Poor Rate. Usually this took the form of offering premiums to those families who kept themselves off the parish rates, and occasionally premiums were offered to the person who wrote

the most clear, comprehensive, and satisfactory Essay or Treatise for a Reform of

[4] J. R. Poynter, *Society and Pauperism* (London: Routledge and Kegan Paul, 1969), pp. 19 and 188.

[5] 'I had rather,' he wrote at a later date, 'that all the commons of England were sunk in the sea, than that the poor should in future be treated on enclosing as they have been hitherto' (quoted in Hobsbawm, *Industry and Empire*, p. 82).

[6] J. Townsend, *A Dissertation on the Poor Laws* (London: C. Dilly in the Poultry, 1786), in J. R. McCulloch (ed.), *A Select Collection of Scarce and Valuable Economical Tracts . . .* (London: printed by Lord Overstone, 1859), pp. 411–12. For similar discussions *see* James Anderson, 'To the Right Honourable Lord Sheffield', in James Anderson (ed.), *Recreations in Agriculture, Natural-History, Arts, and Miscellaneous Literature*, VI (1802), 38; A.Z., 'Obstacles to Agricultural Improvement', *The Farmer's Magazine*, III (1802), 305–6; and Thomas Thomson, *History of the Royal Society* (London: Robert Baldwin, 1812), p. 64.

the Poor Laws, in this country, shewing by what practicable and easy method the poor may be generally enabled to escape the necessity of applying to the parish for relief . . . .[7]

Nor was the Speenhamland system of 1795, which regulated the Poor Rate by the price of corn, very successful, and many British historians consider it 'a shorthand expression for rural pauperization'.[8] It was thus a vicious circle (or epicycle): enclosure and improvement, which were probable causes of poverty, were supposed to relieve the problems of food shortage and misery; but these very problems made such improvement unlikely. Philanthropy and agriculture were inextricably intertwined; if the Royal Institution began as a philanthropic organization, we should not be surprised at its subsequent agricultural orientation.

Improvement and enclosure were, of course, methods of indirect philanthropy (if they were charitable at all). After 1795, with the excesses of the French Revolution hanging over it, and in a climate of food riots and incipient rebellion, the aristocracy cast about for more immediate solutions. 'You can not feel the dangerous state of the country stronger than I do,' wrote the Earl of Egremont to Arthur Young, Secretary of the Board of Agriculture, in 1797.

> [O]ur present situation has not come on unexpected by me altho[ugh] it may be by our Ministers who I suppose do not yet expect to be hanged which they are certain to be & I suppose will not begin to think of preserving their own lives till they feel the rope about their necks.[9]

For Egremont and other members of the Board of Agriculture and the SBCP, allotment offered a direct method to stave off such a development. The idea of the programme was to keep the labourer from becoming dependent on prices at the shop. If the great and lesser landowners would set their tenants and cottagers up with some means of support—a plot of land, a cow, or some cottage industry—then the latter would be able to live an independent life. Young believed that such a programme would strengthen the political loyalties of the poor; some believed it would reconcile the poor to enclosures; and all of those involved hoped it would lower the Poor Rate.[10] Finally, allotment was the type of assistance the landlord was particularly suited to give. Most industry at this time was what is called 'bye-industry' or the 'putting-out' system, and the

---

[7] *Rules, Orders and Premiums, of the Bath and West of England Society, for the Encouragement of Agriculture, Arts, Manufactures, and Commerce* (Bath: R. Cruttwell, 1797), p. 53. An examination of any of the premium lists of the agricultural societies of Durham, Kent, Cardigan, Bath, Manchester, West Riding, Sussex, Norfolk, etc., always reveals mention of an award to the labourer who raises the largest family without depending on the parish allowance, or some variety of the same award.

[8] Asa Briggs, *The Making of Modern England* (New York: Harper and Row, 1965), p. 59. For a dissenting view see Mark Blaug, 'The Myth of the Old Poor Law and the Making of the New', *The Journal of Economic History*, XXIII (1963), 151–84, and 'The Poor Law Report Reexamined', *The Journal of Economic History*, XXIV (1964), 229–45.

[9] BM Add. MS. 35127, f. 407.

[10] Poynter, *Society and Pauperism*, pp. 99 and 103; SBCP *Reports*, I, 42–3; and B. Kirkman Gray, *A History of English Philanthropy* (London: P. S. King and Son, 1905), p. 239.

typical worker was the cottage worker, unaware that he was turning from peasant to wage labourer. Cottage industries were closely connected with agriculture; as late as 1811, more than 35% of the total national income was produced by this sector of the economy.[11] If the landlord set up a tannery, the hides were from his cattle and the bark used in the tanning process from his trees. If his cottagers wove yarn on a loom to produce piece goods, the wool came from his sheep. For various reasons, however, the movement never really caught on. Advocates of allotment, like improving landlords (often, as we have noted, the same people), were the exception rather than the rule.[12]

The overlap in membership among 'improvers' has to be kept in mind if we are to understand the social causes behind the foundation of the Royal Institution. In a climate of fear of revolution and lucrative opportunities for farming, the Board of Agriculture was formed in 1793 to spread the zeal for enclosure and improvement among the landowning class. As social tensions mounted and rioting became increasingly frequent, allotment seemed a more direct answer, and a separate organization, the SBCP, was formed in 1796 to advocate this plan of action. The final step in this sequence was the creation of the Royal Institution in 1799, for reasons that will be discussed shortly. Despite the separateness of these organizations, however, their leaders and most active members were the same people. The core group of the SBCP, for example, roughly duplicates the governors and most active members of the RI. This includes the SBCP President, the Bishop of Durham, who coined the slogan of the allotment movement, 'three acres and a cow',[13] which became 'the Board [of Agriculture's] panacea for the ills of rural society'.[14] It also includes the first President of the Royal Institution, the Earl of Winchilsea, whose name 'recurs so inevitably in every allusion to the subject as to create a suspicion that the movement and his estates were coextensive'.[15] Under what became known as the 'Winchilsea system', a cottager was given between one and four cows, in addition to a plot of land for cultivation.[16] Years later Sir Thomas Bernard, a founder of the RI and editor of the SBCP *Reports*, reported that not a single

[11] Phyllis Deane and W. A. Cole, *British Economic Growth 1688–1959* (2nd ed.; Cambridge University Press, 1967), p. 161. On the relationship between industry and agriculture *see* Mantoux, *Industrial Revolution*, pp. 251 and 474–7; Gray, *English Philanthropy*, p. 240; and T. S. Ashton, *The Industrial Revolution 1760–1830* (New York: Oxford University Press, 1964), p. 25 and Ch. 3.

[12] On allotment *see* Poynter, *Society and Pauperism*, pp. xviii-xix, 98–103, and 163–4; the Hammonds, *Village Labourer*, pp. 150–7; and Chambers and Mingay, *Agricultural Revolution, passim*. For a contemporary account *see* the several volumes of the *Communications to the Board of Agriculture*, which reprinted many articles published in the SBCP *Reports*. *AY* from 1795 onward also contained much material on the subject.

[13] Timothy Eden, *Durham* (2 vols.; London: Robert Hale, 1952), II, 382.

[14] Winifred Harrison, *The Board of Agriculture, 1793–1822, with Special Reference to Sir John Sinclair* (M.A. Dissertation, University of London, 1955), p. 83. I am grateful to Mrs Winifred Harrison Matthews for permission to quote from her thesis.

[15] The Hammonds, *Village Labourer*, p. 151.

[16] The Eart of Winchilsea (George Finch), 'Extract from an account of cottagers renting land', SBCP *Reports*, I, 108. A good description of Winchilsea's plans for allotment reform is quoted from his own proposals of 1797 in W. Hasbach, *A History of the English Agricultural Labourer*, trans. Ruth Kenyon (London: P. S. King and Son, 1908), pp. 165–8.

farmer on the Earl's estates was forced to go onto the Poor Rate.[17] Another key figure is Sir John Sinclair, first President of the Board of Agriculture.[18] Sinclair, the Dukes of Bedford and Bridgewater, the Earl of Egremont, and many other early RI Proprietors were involved in allotment projects, and the SBCP *Reports* include articles on allotment by Bernard, Winchilsea, Samuel Glasse, Rowland Burdon, Edward Parry, William Morton Pitt, Lord Teignmouth, and the Bishop of Durham. No less than eight of the first nineteen governors of the RI were either directly involved in allotment or making studies of allotment experiments.[19]

If the Board of Agriculture was trying to mobilize interest in enclosure and improvement, and the SBCP in allotment, why found yet another institution to combat rural poverty? How could the Royal Institution hope to add to these campaigns? The answer to this lies once again in the dual revolution. Specifically, the French Revolution generated an interest in educational organizations for the poor as a means of instilling appropriate political attitudes, while the Industrial Revolution provided a definition of science that seemed quite applicable to the relief of rural distress. So strong were these tendencies that the creation of an institution like the RI, which was designed to combine the two, was literally inevitable by the end of the eighteenth century. Count Rumford, of course, was the spark that ignited this mixture; but his role can be understood only if we have the *raisons d'être* of the Royal Institution in clear perspective.

The interest in educating the poor had both religious and scientific aspects to it at this time. The religious influence was derived from the Evangelical revival, particularly the Clapham Sect, many of whose members belonged to the SBCP and later the RI. There was, however, a host of projects popular at this time, as the upper classes tried 'by charital and educational organizations, to instil a sense of order and a respect for property into the working class'.[20] As E. P. Thompson has written, the remedies for social unrest differed,

> but the impulse behind [Patrick] Colquhoun, with his advocacy of more effective police, Hannah More, with her halfpenny tracts and Sunday Schools .... Bishop Barrington's more humane Society for Bettering the Conditions of the Poor, and William Wilberforce and Dr. John Bowdler, with their Society for the Suppression of Vice and Encouragement of Religion, was much the same .... 'Patience, labour,

[17] Thomas Bernard, *A Letter to the Honourable and Right Reverend the Lord Bishop of Durham* ... (2nd ed.; London: J. Hatchard, 1807), p. 21.

[18] John Sinclair was a seminal figure in the movement for agricultural improvement. In addition to establishing an experimental farm, a tannery, and bleachfields on his own estate, and assuming early leadership of the Board of Agriculture, he published a twenty-one volume statistical account of Scottish husbandry. His praise for the Winchilsea system may be found in his 'Observations on the Means of enabling a Cottager to keep a Cow, by the Produce of a small Portion of Arable Land', in *Essays on Miscellaneous Subjects* (London: T. Cadell, Jr. and W. Davies, 1802), p. 31.

[19] *See* below, pp. 41ff.

[20] Betsy Rodgers, *Cloak of Charity* (London: Methuen, 1949), p. 12.

sobriety, frugality and religion, should be recommended to [the poor]; all the rest is downright fraud.'[21]

Religious education was a popular theme in the SBCP *Reports*, and Thomas Bernard pointed out that it has 'been a primary and unvaried object of our Society ...'.[22] In general, the SBCP championed education as a means of mitigating social unrest. Thus Hannah More's sister wrote in 1800 that the Bishop of Durham 'fully feels the importance of instructing the poor, as the grand means of saving the nation'.[23] Similarly Bernard wrote in the *Reports*:

> The absurd prejudices that *have* existed against extending the common and general benefits of education to the children of the poor, and the extraordinary supposition, that an uneducated and neglected boy will prove an honest and useful man,—that a youth of ignorance and idleness will produce a mature age of industry and virtue,—are now in great measure exploded.[24]

The scientific aspect of such education was seen as a natural partner of religion, for both disciplines placed great emphasis on the 'natural order of things', both in and out of their respective spheres. As the contemporary interest in natural theology indicates, science and religion were, at least at this point in time, comfortable bedfellows, and the Reverend Paley popularized the notion of hierarchy in nature as a scientific verification of hierarchy in society.[25] Davy's first chemistry lecture at the Royal Institution included this as a theme. The man who is familiar with science, he told his audience, will perceive in the universe 'the design of a perfect intelligence ... will be averse to the turbulence and passion of hasty innovations, and will uniformly appear as the friend of tranquillity and order'.[26] A place where the poor could be educated in such principles was thus high on the list of SBCP priorities.

In addition to this sublime and heuristic value of science, there seemed to be a more practical task to which the scientific method could be put. There was, in the eyes of SBCP members, such a thing as 'scientific' philanthropy, by which

---

[21] E. P. Thompson, *The Making of the English Working Class* (Harmondsworth: Penguin Books, 1968), pp. 60–1. The quotation is from Edmund Burke, 1795. On the Evangelicals the most recent study is Ford K. Brown, *Fathers of the Victorians* (Cambridge University Press, 1961). *See also* David Spring, 'The Clapham Sect: Some Social and Political Aspects', *Victorian Studies*, V (1962), 35–48. There are also a number of biographies of Claphamites who were in the RI, among them Standish Meacham, *Henry Thornton of Clapham* (Cambridge: Harvard University Press, 1964); Lord Teignmouth, *Memoir of the Life and Correspondence of John Lord Teignmouth* (2 vols.; London: Hatchard and Son, 1843); and several portraits of William Wilberforce.

[22] *A Letter to ... the Lord Bishop of Durham ...*, p. 21. *See also* SBCP *Reports*, II, 213–14; III, 1–41; and IV, 1–42.

[23] William Roberts, *Memoirs of the Life and Correspondence of Mrs. Hannah More* (2 vols.; New York: Harper and Brothers, 1855), II, 59.

[24] SBCP *Reports*, II, 193, dated 7 October 1799.

[25] An idea that experienced something of a revival in the 1960s in animal research on territory and aggression, notably by Konrad Lorenz, Desmond Morris, Robert Ardrey, *et al.* On natural theology in pre-Darwinian England *see* Charles Gillispie, *Genesis and Geology* (New York: Harper and Row, 1959), Ch. 8.

[26] *DW*, II, 326.

term they meant charity that was 'systematic' or 'organized'.[27] 'Let us therefore make the inquiry into all that concerns the *poor*, and the promotion of their happiness, a *science*' wrote Thomas Bernard; 'let us investigate *practically*, and upon *system*, the nature and consequences of those things which experience hath ascertained to be beneficial to the poor.'[28] Much of this was derived from the ideas of Jeremy Bentham, whose ideology of science came to play a crucial role at the RI in years to come. A man like Patrick Colquhoun, a member of the SBCP who helped plan the formation of the RI, is a good example of Benthamism come to philanthropy. To Colquhoun, the scientific study of poverty meant compiling data on the distribution of wealth, or on the demography of thieves and prostitutes, as well as establishing a police network.[29] It also meant efficiency in the kitchen and the cottage, and Colquhoun and other members of the SBCP were soon preoccupied with setting up soup kitchens and inventing new recipes for the poor. 'An infinite vista of kitchen reform beckoned to their ingenious imaginations,' wrote the Hammonds.[30] This had natural connections with the allotment movement, for scientific philanthropy was easily extended to include various suggestions on how to build cottages cheaply, what were the best types of cooking utensils, how the poor could heat their cottages with a minimum of fuel, etc. It was this supposedly scientific treatment of scarcity that brought Benjamin Thompson, Count Rumford, the most important foreign influence on English philanthropy during this period, onto the British scene.[31] In 1795, when Thomas Bernard became Treasurer of the Foundling Hospital, he instituted a system of fuel and food economy based on Rumford's ideas. Bernard has quite rightly been called the 'man who was second only to Colquhoun in spreading the Rumford gospel in England . . .'.[32] In 1797, Rumford was made a member of the SBCP and its General Committee for life, 'in consideration of [his] extraordinary services . . . for the benefit of the poor . . .'.[33] The SBCP *Reports* regularly featured articles on how Rumford's 'scientific' principles on food and fuel were adopted for use in schools, hospitals, and chimneys, and in 1798 the Bishop of Durham pointed

[27] David Owen, *English Philanthropy 1660–1960* (London: Oxford University Press, 1965), p. 105, uses the term 'associated philanthropy' to describe the organization of charity into formal groups.

[28] SBCP *Reports*, I, xii; italics in original. *See* also *A Letter to . . . the Lord Bishop of Durham* . . ., p. 59.

[29] Besides works by Colquhoun, such as his *Treatise on Indigence* (1806), there is a sketch in the *DNB*, XI (1887), 403–5, by Francis Espinasse, and one by Oscar Sherwin, 'An Eighteenth Century Beveridge Planner', in *The American Historical Review*, LII (1947), 281–90. Material can also be found in Owen, *English Philanthropy*, pp. 100–2 and 108; William Allen, *Life of William Allen* (3 vols.; London: Charles Gilpin, 1846), I, 33ff.; and Leon Radzinowicz, *A History of English Criminal Law and Its Administration from 1750* (4 vols.; London: Stevens and Sons, 1948–68), III, 211ff. Finally, the reader should consult the posthumous memoir by Iatros [Grant D. Yates], *A Biographical Sketch of the Life and Writings of P. Colquhoun, Esq.* (London, 1818).

[30] The Hammonds, *Village Labourer*, p. 119.

[31] Poynter, *Society and Pauperism*, p. 87.

[32] Fritz Redlich, 'Science and Charity: Count Rumford and His Followers', *International Review of Social History*, XVI (1971), 202.

[33] SBCP *Reports*, I, 286–7.

1a. Benjamin Thompson, Count Rumford

1b. Thomas Garnett

1c. Thomas Young

to the 'variety of useful and extraordinary inventions and improvements, which Count Rumford has made, and *is now making*, for the benefit of mankind'.[34] Jeremy Bentham enthusiastically pictured a system of workhouses outfitted with Rumford stoves and utensils and having the Count as its administrator.[35] To the English upper class, he represented the essence of scientific philanthropy.

Nevertheless, it was Rumford's timing rather than his talent that brought the Royal Institution into being. The climate of fear and hysteria during the closing years of the eighteenth century is well documented, and Rumford's biographer, George Ellis, was correct in pointing out its importance to the Count's success.[36] His competence, real or imagined, was thus much beside the point, for he appeared to offer the possibility of alleviating the misery of the poor (and therefore the fear of the rich) without even a minor alteration in English social structure. The years 1796–7, and 1799–1800, were peaks in the development of soup charities. Rumford's arrival in England in 1798, and his proposal of an institution for relieving the distress of the lower classes by spreading practical scientific knowledge and introducing practical scientific improvements, clearly came as something of a godsend. Given the context, Rumford's role in the foundation of the RI seems almost arbitrary. In order to understand the contingent nature of his relationship to the Institution, it will be necessary to review some of his personal history prior to 1799, and carefully constructed reputation among the British aristocracy.[37]

[34] ibid., II, 42. *See* I, 33, 65–71, and 263; and II, 31–42, 231–6, and 243.

[35] Poynter, *Society and Pauperism*, p. 133.

[36] George E. Ellis, *Memoir of Sir Benjamin Thompson, Count Rumford* (Boston: American Academy of Sciences, 1871), p. 191.

[37] The following sketch of Rumford does not pretend to exhaust the subject of this complex figure, but represents what I regard as the most plausible portrait given the available evidence. The more one reads in and about Rumford, the less one is tempted to take him seriously. Although he was a keenly ambitious and industrious man, his ideas tend to be typical of his era, and never profound. As *The Diary of Sir Charles Blagden* (eight MS volumes in the archives of the Royal Society) reveals, Rumford's immense prestige was strangely mirrored by widespread private mockery, both in England and on the Continent; and this sentiment may have been justified. He was, in many ways, a case study in the emperor's new clothes. Although the diary of the RS Secretary has a gossipy quality to it, I have found it a reliable testimony, corroborated by the few other unpublished materials available, as well as by some of George Ellis' inadvertent remarks in his biography of Rumford. The diary was a private affair, not intended for publication or for influencing public opinion. It has no prose style, being a grammarless set of notes made by Blagden to himself, and there would have been no point in recording false information or non-existent testimony. Although he personally disliked Rumford, Blagden was, like Sir Joseph Banks and so many others, publicly on good terms with the Count, and the diary records this dual relationship Rumford had with the English aristocracy. As for Rumford's role in the early history of the RI, I have presented what I consider to be the most likely balance among conflicting testimony, but this has made it necessary to omit a good bit of detail on Rumford which would, if included, constitute a lengthy digression from the story of the RI itself. (This material can be found, however, in the set of appendices to this work, deposited at the Royal Institution and The Johns Hopkins University, as explained in the Preface.)

Standard biographical information on Rumford, in addition to Ellis, can be found in James A. Thompson, *Count Rumford of Massachusetts* (New York: Farrar and Rinehart, 1935); Egon Larson [E. Lehrburger], *An American in Europe* (New York: The Philosophical Library, 1953); Sanborn C. Brown, *Count Rumford: Physicist Extraordinary* (Garden City: Doubleday, 1962); W. J. Sparrow, *Knight of the White Eagle* (London: Hutchinson, 1964); and Duane Bradley,

The most consistent feature of Rumford's personality was his inability to commit himself to a project for any length of time. This was not just a problem of shallowness or opportunism, for Rumford so often deserted the situations he created that he left an enduring trail of deep resentment behind him. Rumford was born in Woburn, Massachusetts, in 1753, studied medicine, and worked as a schoolteacher until he married a wealthy widow in 1772. 'The material advantages Thompson gained from the union,' writes one biographer, 'were not meagre.'[38] His father-in-law, one Colonel Walker, was one of the leading figures in Concord, New Hampshire, and it was through Walker and his daughter that Rumford met John Wentworth, the governor of the province, from whom he received a commission as Major in the Second Provincial Regiment. At the outbreak of the Revolution Rumford escaped from the Colonies as a Royalist, having been employed as a spy by the British forces, and left his wife and infant daughter, Sarah, behind.[39] In England, he formed a liaison with Lord George Germain, the Secretary of State for America, who appointed him Secretary to the Province of Georgia and (in 1780) Under-Secretary of State for the Northern Department. By 1781 Rumford also held the posts of inspector of all clothing sent to America, and Lt.-Col. Commandant of Horse Dragoons at New York, reputedly drawing an annual income of nearly £7,000 from his collective appointments.[40] Soon after, however, he was accused of embezzlement and treason—selling naval information to the French—and Germain found it necessary to reassign him to America.[41] In 1783, travelling on the Continent, he made the acquaintance of the Elector of Bavaria, and was later taken into his service. Rumford instituted various reforms in the Elector's army,[42] and made a name for himself by using the military to round up the beggars of Munich in 1790 and put them into workhouses. Resentment, however, developed between Rumford, the Munich City Council, and the army, and in 1793 he set out for Italy. There he met a number of the British aristocracy, including the Palmerston family and Sir Charles Blagden, Secretary of the Royal Society. Lady Palmerston's interest in Rumford, in particular, assisted his entry into British society.[43] After a short return to Munich, Rumford obtained a

Count Rumford (Princeton: D. Van Nostrand, 1967). *BJ*, of course, reproduces Rumford's account of his role in the formation of the RI, but it should be noted that this is taken from Rumford's own version in *The Complete Works of Count Rumford* (4 vols.; Boston: American Academy of Arts and Sciences, 1875), IV, 747–54. Useful articles include T. E. James, 'Rumford and the Royal Institution: A Retrospect', *Nature*, CXXVIII (1931), 476–81, and A. H. Ewen, 'The Friend of Mankind: A Portrait of Count Rumford', *Proceedings of the Royal Institution of Great Britain*, XL (1964), 186–200.

[38] Sparrow, *White Eagle*, p. 23. In general I shall refer to Thompson as 'Rumford' even though he did not receive his title from the Elector of Bavaria until 1791.

[39] Alan Valentine, *Lord George Germain* (Oxford: Clarendon Press, 1962), p. 462, and Sparrow, *White Eagle*, Ch. 2.

[40] *BJ*, pp. 17–19.

[41] Charles R. King (ed.), *The Life and Correspondence of Rufus King* (6 vols.; New York: G. P. Putnam's Sons, 1896), III, 518–19; Valentine, *Germain*, p. 474.

[42] The reforms collapsed upon his departure. Redlich ('Science and Charity', p. 185) regards Rumford as a very poor army reformer, and argues that the Bavarian experiment was a dismal failure.

[43] This relationship is described in detail in Brian Connell, *Portrait of a Whig Peer* (London:

leave of absence from the Elector, and arrived in England late in 1795. Here he remained until the spring, travelling to Dublin and Edinburgh to make various improvements at local institutions, and then, after a short stay in London, returned to Bavaria. Rumford was back in England in September of 1798, when he became involved with the SBCP and their plans for what would become the Royal Institution.

Rumford's real career consisted of being something of a professional courtier, and he had the ability to advertise his own talents to advantage. He was, as one Austrian general so aptly put it, 'the hero of his own panegyric',[44] an opinion confirmed repeatedly by Blagden in his diary. According to Ellis, Rumford 'craved intercourse on confidential terms with many of the nobility',[45] and his interest in scientific subjects, especially those related to current problems, assured him a favourable reception. In 1796, Rumford donated £1,000 to the Royal Society in stocks for an annual medal to be awarded in his name, for studies on heat. Between 1796 and 1800 his *Essays, Political, Economical and Philosophical* began to appear, covering such topics as the design of chimneys and the economy of fuel, which were soon popularized by other writers.[46] F. M. Eden's very influential *State of the Poor* (1797), for example, provided lengthy quotations from the *Essays*. In 1796 Rumford donated 'many ingenious models of useful machines' to the Board of Agriculture, and attempted to improve the chimney of the Board building on Sackville Street.[47] His introduction to the British nobility, via the Palmerstons, resulted in more than fifty aristocratic families having him install stoves or reconstruct chimneys in their homes.[48] Rumford also redesigned the chimney at Sir Joseph Banks' house, and the kitchen at Sir John Sinclair's. The latter contained one of his fuel-saving hearths, and Sinclair kept the kitchen open to public inspection at all hours of the day.[49]

Rumford's pursuit of the aristocracy was, however, not enough to ensure him any permanent respect or allegiance. As in the case of his previous involvements, he was vilified by precisely this group soon after his departure in 1802. Had the Royal Institution simply been the product of his interests or personality, it would never have been founded, let alone have lasted beyond his departure; and in fact, Rumford was planning to return to America in 1799. Furthermore, his ideas on fuel and food may not have been sound, and much of what he had written on poverty was banal, even in contemporary terms.[50] But in 1798, the need for such an institution like the RI was very clear, and Rumford's immense, if transient, prestige proved to be the necessary catalyst.

Andre Deutsch, 1957). Rumford wrote her sixty-six letters between 1793 and 1804.
[44] Quoted in Sparrow, *White Eagle*, p. 105.
[45] Ellis, *Rumford*, p. 232.
[46] T. Danforth, *The Theory of Chimnies and Fire-Places investigated; the principle of those recommended by Count Rumford fully explained* ... (1796); D. Braidwood and Son, *Count Rumford's chimney fire-places improved, containing the outlines of his plan* ... (1797?).
[47] Ellis, *Rumford*, pp. 204 and 233–4; *see also* Redlich, 'Science and Charity', p. 198.
[48] Ewen, 'Friend of Mankind', p. 192.
[49] Ellis, *Rumford*, pp. 233–4 and 276–7.
[50] Poynter, *Society and Pauperism*, p. 89.

The earliest mention of what would become the Royal Institution is contained in a set of 'Proposals' by Rumford which appeared in 1796.[51] Rumford's description of the Institution was typical of voluntary organizations of the time, both in its administration by the wealthy and in its sponsorship of soup kitchens on the premises. It also added a notion, that of a 'House of Industry' or public display of inventions, taken from ideas then current in France and Scotland.[52] Rumford was also very specific in assigning himself a central role in its realization, a commitment which proved to be a major source of discord in the early administration of the RI. 'The general arrangement of the establishment and of all its details,' he wrote, 'will be left to the author of these proposals, who will be responsible for their success.'[53] In short, Rumford presented himself as an expert in search of a patron.

Rumford began his correspondence with Thomas Bernard soon after being made a life member of the SBCP. Writing from Munich on 8 June 1798, he stated:

> A well-arranged House of Industry is much wanted in London .... Pray read once more the 'Proposals', published in my second Essay. I really think that a public establishment, like that there described, might easily be formed in London, and that it would produce infinite good. I will come to London to assist you in its execution whenever you will in earnest undertake it.[54]

For his part, Bernard was enthusiastic. 'This gentleman I found,' wrote Rumford,

> on my return to England in September 1798, not only agreeing with me in opinion in regard to the utility and importance of the plan I had proposed, but very solicitous that some attempts should be made to carry it into immediate execution in this capital.[55]

[51] *Essays, Political, Economical and Philosophical* (4 vols.; London: T. Cadell, Jr., and W. Davies, 1796–1812), I, 113–88. These were printed up separately by Cadell and Davies the same year. *See also The Complete Works of Count Rumford*, IV, 327–93, esp. pp. 361–9. The full title is: 'Proposals for forming, by private subscription, an establishment for feeding the Poor, and giving them useful Employment; And also for furnishing Food at a cheap Rate to others who may stand in need of such Assistance. Connected with an Institution for introducing, and bringing forward into general Use, new Inventions and Improvements, particularly such as relate to the Management of *Heat* and the Saving of *Fuel*; and to various other mechanical Contrivances by which *Domestic Comfort* and *Economy* may be promoted'.

[52] Rumford was well aware of the existence of the Conservatoire des Arts et Métiers (1794) in Paris, and several authors have followed Elie Halévy (*History of the English People in the Nineteenth Century*. Vol. I: *England in 1815*, trans. E. I. Watkin and D. A. Barker [2nd ed.; London: Ernest Benn, 1961 reprint], p. 566) in asserting that the RI was a deliberate copy. Thomas Kelly, in *A History of Adult Education in Great Britain* (Liverpool: University Press, 1963), p. 103, states that the RI was derived from Anderson's Institution in Glasgow (1796), and Vernon ('Foundation and Early Years of the RI') tends to agree. Thomas Garnett, the RI's first professor, taught at Anderson's before coming to the RI, and describes it in his *Observations on a Tour Through the Highlands* (2 vols.; London: 1800), II, 193–205. It also had instruction for working men, a library, a repository of inventions, a laboratory and a workshop. Its lecture theatre was very similar to the one at the RI, and it is probably not irrelevent that the architect of the latter, Thomas Webster, was a Scot.

[53] Rumford, *Works*, IV, 363.

[54] ibid., p. 749n.

[55] ibid., p. 749.

The two men met several times after that, in consultation with the Bishop of Durham, William Wilberforce, and others. The SBCP finally appointed a Select Committee of eight men to meet with Rumford and review the proposal, and they did so on 31 January 1799, at the home of Richard Sulivan. The following day, the Committee reported its satisfaction with the final draft, and Rumford sent them a corrected copy on February 7th. 'I am perfectly ready,' he wrote in the accompanying letter, 'to take any share of the business of carrying the scheme into execution . . . .'[56]

This new set of 'Proposals' was more specific than the essay of 1796. The repository would contain cottage fireplaces and kitchen utensils; a model kitchen for a farmhouse; stoves, grates, boilers, and kettles; models of the steam engine, as well as of ventilators, hothouses, agricultural machinery, lime kilns, bridges, and cottages; farm implements; and spinning-wheels. This collection, Rumford stated, would provide for the diffusion of new inventions and ideas in domestic economy, and any subscriber would be able to commission drawings of these models at his own expense. Lord Winchilsea, who was to be the first President of the RI, saw this as vital to the allotment programme, and wrote to Arthur Young how impatient he was for the time when 'these things are shewn in use at the Institution'.

> I wish with you [he continued] that all this could have been done before this time, the People here are much distressed for Fuel, in many Parishes none is to be bought at any price . . . .[57]

As for the teaching function of the institution, this would be accomplished via a lecture room and laboratory. Rumford especially recommended lectures on heat, fuel, combustion, clothing, ventilation, refrigeration, vegetation, manures, digestion, tanning, soap-making, bleaching, and dyeing. The rest of the 'Proposals' dealt with the subscriptions and finances of the institution, and the nature of its government and administration. The key contributors, called 'Proprietors', would pay fifty guineas—a very large sum in 1799—for the privilege of hereditary membership.[58]

These new 'Proposals' were subsequently circulated among the wealthy classes in London, and received a quick response from fifty-eight of Britain's most famous families. Some of this group—how many is not clear—then met at the home of Sir Joseph Banks on 7 March 1799 to found the institution, and elect the Committee of Managers (at first nine men), which had been chosen in

---

[56] ibid., p. 751.

[57] BM Add. MS. 35128, ff. 172–3; dated 6 January 1800.

[58] Rumford's 'Proposals for forming by Subscription, in the Metropolis of the British Empire, a Public Institution for diffusing the knowledge and facilitating the general introduction of useful mechanical inventions and improvements, and for teaching by courses of philosophical lectures and experiments the application of science to the common purposes of Life' are reprinted in *BJ*, pp. 121–34, and in Rumford, *Works*, IV, 755–64.

The 50-guinea membership is a good indication of the social class we are dealing with. At this time £25 was the average annual wage for unskilled labour, and £100 a comfortable middle-class income.

advance by the Select Committee of the SBCP. On March 9th, the Managers
assembled at Banks' house once more, at which time they planned to have 500
copies of Rumford's 'Proposals' printed, solicit further subscriptions, and present a copy of the 'Proposals' to George III. The King became a patron in June,
giving the Institution the right to call itself 'Royal'. In September, Bernard
arranged the purchase of the mansion of a John Mellish at 21 Albemarle Street,
for £4,850, and a Clerk of the Works, Thomas Webster, was engaged. The
Royal Institution was well on its way.[59]

Within a few weeks after the first meeting, however, Rumford's role in the
running of the Institution became confused. He had made it clear to Bernard
and the SBCP that he would shoulder most of the responsibility involved in
organizing the RI, yet he had been planning, as early as 1797, to return to
America in the spring of 1799. He reaffirmed this intention in a letter to his
friend, Loammi Baldwin, on 28 September 1798, and wrote to his publisher on
26 January 1799 that he would be setting sail in March.[60] Not until the middle
of March did Rumford reverse his plans. In a letter to Rufus King, an American
diplomat living in England, he enclosed a copy of the Institution's 'Prospectus'
and gave it as his reason for deciding to remain in the U.K.[61] Even then, he
wrote to his mother on March 18th, that his stay in England would probably be
only about a year in length.[62]

Until mid-March, then, Rumford had led Bernard, Banks, and the rest of the
core group at the RI to believe that he was seriously committed to the project,
and the subsequent duplicity had strong repercussions for the management of
the Institution. This pattern, so typical of his entire life style, turned Rumford
into something of a titular figure overnight. The real movers behind the RI had
to have his presence and reputation if the venture were to succeed; but they
found it necessary to render his real position innocuous, a situation which a
man as egocentric as Rumford, so long as he *had* to be involved, would naturally resent. This inherent ambiguity is the reason, I believe, for the common mis-
interpretation of Rumford as the founder and central figure of the Royal
Institution. He was in charge of building construction, workmen and personnel,
and some of the correspondence, since the governors deliberately wished the
place to be identified with his name; yet in reality, he was very nearly
superfluous. In America, in Bavaria, now in England, and later in France, there
were always two Rumfords operating.

Sir Joseph Banks' annoyance with Rumford began when the latter simply
neglected to do any work in raising subscriptions. By the end of March, Banks

[59] *BJ*, pp. 135–7; Rumford, *Works*, IV, 765–8; *MM*, I, 55–6 (14 September 1799). Bernard
also prepared the RI Charter; *see* his autobiography, *Pleasure and Pain*, ed. J. Bernard Baker
(London: John Murray, 1830), p. 59.

[60] The 1799 letter is in BM Add. MS. 34045; *see also* Sparrow, *White Eagle*, p. 152. On Rum-
ford's earlier plans *see* Ellis, *Rumford*, pp. 342–3, and the Earl of Ilchester (ed.), *The Journal of
Elizabeth Lady Holland* (2 vols.; London: Longmans Green, 1908), I, 207 (entry for 20
November 1798).

[61] Ellis, *Rumford*, p. 353; *BJ*, pp. 64–5.

[62] Letter of 18 March 1799, in the private collection of William C. Pierce of New York City. A
copy of this is preserved at the RI.

was fast losing patience with the Count. Blagden visited Sir Joseph on 29 March 1799 and recorded the following:

> [T]alk about C[oun]t Rumford: his management Sir Joseph thought scarcely fair: I said I had never seen anything unfair: [Banks said] Subscriptions not coming in: those already obtained, chiefly by dinners and management [i.e. by the Managers]. Sir Jos. began to think with me that probably C[oun]t Rumford had better to have gone to America for himself.[63]

Shortly after this Rumford indicated his desire to Banks to withdraw himself from any future role in the Institution and have the latter take over as President. Blagden records that Banks' reaction was one of justifiable anger:

> Banks told him he had yoked himself with 8 asses, & now he wanted to get out of the harness, and turned carter [traitor]. I told [Banks] that I certainly had less confidence in C[t] R[umford] than I formerly had .... Banks said he co[d.] [could] have no confidence after the attempt made to engage him [Banks] to be President. [T]hat if he piqued himself on anything it was on keeping true to his old engagements.[64]

Blagden's testimony is corroborated by a letter Banks wrote to the Earl Spencer less than four weeks later. It reveals that Institution affairs had picked up financially and membership was increasing; and it also shows that Rumford's desire to abandon the venture to Banks and others had been realized, although not publicly. Finally the letter reveals that Rumford had lost his standing with the core group at the RI, who no longer took him seriously.

Soho Square
May 6 1799

My dear Lord

    We have now Final[l]y arrang[e]d (I believe) our Institution matters Lord Bessborough who is one of the Visitors is to change Places with your Lordship & L[or]d Winchilsea to be chosen a manager in order that he may be our President which his Lordship has undertaken to be I am told with some symptoms of Satisfaction.

    ... Every thing yet done by the managers has been approved warmly by the Proprietors at their Last meeting and another meeting must soon be Call[e]d to Elect L[or]d Winchilsea an additional manager ... so well do we prosper in Point of Finance as well as of increasing members that we hope in a very short time to have enough to pay the Purchase of the house ....

    Count Rumford has of late kept himself intirely [*sic*] in the background nor do I think he will ever venture forward again [;] it is [?] has cost him a Fit of sickness to find his way from the Ideal preeminence of his Character to his actual situation in the sciences of this country but I believe he has now satisfied himself & will be as he ought to be an extremely usefull [*sic*] inventor of machines & governor of machine makers

I have the honor to be
With infinite Regard & Esteem
your Lordships Obedient
H[um]ble Ser[vant]
Jos: Banks[65]

[63] *The Diary of Sir Charles Blagden* (*see* footnote 37, above), III, 29 March 1799.

[64] ibid., 12 April 1799. 'Turned carter' is an expression probably adopted from the story of Richard Carter, an admiral who was rumoured to have been paid £10,000 by the French to lose a battle deliberately in 1692.

[65] This letter is only partially reproduced in Warren R. Dawson (ed.), *Supplementary Letters*

By September, others of the nobility whose favour he had successfully courted, e.g. Lord Palmerston, 'agreed that C¹ Rumfords a bad scheme, [and] that [he would have done] better to have gone to America'.[66] The picture that emerges of the Royal Institution is not one of a philanthropic institution that changed its purpose upon Rumford's departure in 1802, but rather of one that was out of his hands almost within its first month. It is thus not surprising that he was absent, between March 1799 and April 1802 (when he formally severed his connection), nearly 40% of the time. Rumford did direct some of the activity at the RI, and did hold some of the account books; but he was regarded by Banks, Spencer, Winchilsea, Bessborough and others as a drawing card without the Institution and a 'governor of machine makers' within it. His occasional presence at Albemarle Street was marked by heavy conflict with the Managers and Visitors, but he was easily pushed aside. His contribution to the RI was over long before the Institution even opened its doors, and his departure in 1802 was marked by bitter feelings. By 1803 he was accused, in absentia, of embezzlement; and as we shall see, the charge may have been justified.[67]

We are now in a position to understand the goals and operations of the Royal Institution during its early years. There were three classes of membership: Annual Subscribers, Life Subscribers, and Proprietors, this last group being the most important. Fifty-eight men became Proprietors right away (although one withdrew shortly after), and thirty-six more joined by April (1799), when the first Committee of Visitors was elected. By May of 1800 the RI had enrolled a total of 280 Proprietors, and between that time and April 1803 it counted another seventy-seven, for a total of 357. By the governors' choice, there were very few additions after that date.

The control of the Institution was in the hands of ten Managers, one of whom also served as President. This Committee made all of the decisions involved in running the Institution, although some responsibility was delegated to Rumford (himself a Manager), Webster, William Savage (the RI printer), Davy, *et al.*, so that the various jobs that had to be accomplished could be done as efficiently as possible. There were also nine Visitors, whose principal func-

---

of Sir Joseph Banks (London: Bulletin of the British Museum [Natural History], Historical Series Vol. 3, No. 2, pp. 41–70, 1962), p. 61. The major part of the deprecation of Rumford is omitted, and I have reproduced most of the original with the kind permission of the present Earl Spencer. The letter is preserved in the Muniment Room at Althorp, Northampton.

The (second) Earl Spencer made a note on the letter that he had replied to Banks, agreeing with the arrangements and sending him £50.

[66] Blagden, *Diary*, III, 30 September 1799.

[67] Rumford was absent for fifteen out of the thirty-six months between March 1799 and April 1802 (*BJ*, pp. 146, 159–61, and 188–9). As for his quarrels with the governors, this was denied by Jones, but affirmed by obituaries of Rumford in *The Monthly Magazine, or British Register* (May 1815) and in Thomas Thompson's *Annals of Philosophy* (April 1815). Blagden records that in 1800 Rumford told him of being outvoted in disputes with Bernard and others (*Diary*, III, 10 March 1800); and Thomas Young's biographer George Peacock, who was old enough to know some of the participants, wrote that the RI's 'founder, if such he may be termed . . . [a]fter managing the affairs of the Institution for a few months . . . quarreled with some of the directors and abandoned the scheme altogether' (*Life of Thomas Young* [London: John Murray, 1855], p. 134).

tion was to submit an annual progress report on the Institution, and especially to review its financial situation. Both Managers and Visitors were elected from the Proprietorship every April and these elections were, if one is to be candid, rigged. The rotational nature of the offices served to confine power to the same set of people, since Managers and Visitors matter-of-factly traded positions from year to year. In other words, within each Committee, there were three men elected to a three-year term, three to a two-year term, and three to a one-year term. Every April, six one-year governors went out of office; the two-year governors became one-year governors; and those with a three-year term dropped down to the two-year slots. Thus six men were elected once a year to fill the three-year offices, and any governor who had gone out of office was immediately eligible for reelection to either Committee. As a result, the RI possessed a strong 'old guard'. Election slates were drawn up by the Managers and never contested. Furthermore, the first Committee of Managers was chosen by the Select Committee of the SBCP, and this Committee in turn chose the first Committee of Visitors the following month (April 1799). The Proprietors simply ratified the Managerial choice. To get some idea of the resulting degree of overlap, when the number of governors was increased in 1803 from nineteen to thirty-one, thirteen of these had been in power in 1799, and by 1806, twelve still remained.[68]

The finances of the Institution pose a more complicated problem, inasmuch as the governors were trying to break new ground. There were no models, in England, of how scientific research and teaching should be supported other than by private initiative and subscription. The Institution was not another version of the Royal Society, and hoped to organize science as a self-sustaining enterprise; but it was difficult to escape past definitions. Once most of the Proprietors and Life Subscribers had enrolled, the major source of extra-Institutional income came from the Annual Subscribers, and this proved to be insufficient. The RI was, in fact, in debt from 1803 to 1836. It struggled to achieve corporate continuance, but at least during its first four decades it survived largely through patronage and occasional, unanticipated benevolence.

There was also an inevitable tension between science for private vs. public use. As we shall see, science at the RI was very much in the service of its Proprietors, who expected certain benefits for their 50-guinea (by 1807, 200-guinea) investment, which were not publicly available. This exclusiveness, however, denied the RI the wide support it needed to remain solvent. The whole question of finances is thus tied to larger questions of the social uses of science, and will have to be discussed in detail at a later point (*see* Chapter 4). For now, it is enough to note that semi-corporate scientific organization proved to be expensive, and that the financial aspects of the project were intimately related to a whole host of new issues which the Managers understandably found difficult, if not impossible, to resolve. The RI did not seem able to find a way to guarantee its own

---

[68] Lords Bridgewater and Palmerston would undoubtedly have been re-elected, but were dead by 1803.

future, and existed for a good part of its history under the chronic threat of dissolution.

What, then, were the functions of the Royal Institution, from 1800, when it opened its doors, to 1802, when Rumford made his official departure? To the early governors, the implementation of scientific philanthropy included a programme of lectures; the publication of a journal; the construction of a Repository, or Model Room, and mechanics' school; and the maintenance of a Laboratory. We discuss each of these in turn.

Private lectures on natural philosophy were available in the metropolis at this time, and were given by men (frequently physicians) such as William Hyde Wollaston in their homes. Groups were small, and largely motivated by the ideals of the gentleman amateur tradition. In the northern provinces, and in Scotland, on the other hand, such lectures tended to focus on technology and what were taken to be the industrial applications of science. It was this new direction which the RI wished to pursue, and thus their first choice for a lecturer was William Farish (1759–1837), then professor of chemistry at Magdalene College, Cambridge.[69] Farish was one of the first in England to lecture on the application of chemistry to arts and manufactures. He made frequent trips through the North to gather information on various mining and manufacturing operations. Farish also possessed what would have been advantageous for the RI's projected Repository, viz. a large collection of models, such as of steam engines, cotton mills, looms, and iron-rolling machines, which he employed in his lectures. Finally, Farish had a commitment to philanthropy, for he was an Evangelical, active in philanthropic institutions, friendly with the Clapham Sect, and rector of St Giles. He was thus an ideal choice.[70]

For whatever reasons, Farish declined to leave Cambridge, and the Managers selected as their second choice Thomas Garnett (1766–1802), a Scottish-trained physician who was then lecturing at Anderson's Institution in Glasgow. This too is not surprising. Anderson's may have been a model for the RI, and Garnett's lectures were known as being practical in nature.[71] His *Outlines of a Course of Lectures on Chemistry* (1797) reiterated the popular theme that chemistry was ancillary to various manufacturing processes, and included material on agriculture, bleaching, dyeing, and calico printing. Like Farish, he also had philanthropic interests, and had set up several charity schemes in Yorkshire.[72]

Rumford wrote to Garnett, requesting him to put his ideas for lectures at the

[69] *MM*, I, 27.

[70] Thomas Cooper, 'William Farish', *DNB*, XVIII (1889), 208; obituary in *Christian Observer*, XXXVII (1837), 611–13, 674–7, and 737–41; J. D. Walsh, 'The Magdalene Evangelicals', *Church Quarterly Review*, CLIX (1958), 499–511. Farish later became Jacksonian Professor of Natural and Experimental Philosophy at Cambridge and first President of the Cambridge Philosophical Society.

[71] *See* above, footnote 52. John Anderson himself, during his lifetime (1726–96), gave lectures on the applications of science to artisans. When Garnett left Anderson's for the RI, he was succeeded by George Birkbeck, the founder of working men's colleges in England, who gave free lectures to mechanics.

[72] Richard Garnett, 'Thomas Garnett, M.D.', *DNB*, XXI (1890), 7–8.

2. 'New Discoveries in Pneumaticks', by Gillray

3. 'The Comforts of a Rumford Stove', by Gillray

4a. Sir John Sinclair

BOARD OF AGRICULTURE.

Apri. 30, 1806.

THE SECRETARY has the Honour to inform you, that Mr. DAVY's Lectures, on the *Application of Chemistry to Agriculture*,

Will commence on Tuesday, ... 6th May,
The Second Lecture will be on Friday
    the .... 9th, —
The Third on Tuesday the ...... 13th, —
The Fourth on Friday the .... 16th, —
The Fifth on Tuesday the .... 20th, —
The Sixth, and last, on Friday the .. 23d, —

      at One
      o'Clock
      precisely.

      N. B. No Lecture will be repeated.

Printed by B. M'Millan,
Bow Street, Covent Garden.

4b. Admission ticket for Davy's lectures before the Board of Agriculture

RI on paper, and Garnett replied on 23 December 1799.[73] He proposed a popular course on experimental philosophy, since he undoubtedly knew, from his own experience, that the subject had its fashionable side and would inevitably draw an audience seeking entertainment. For the more serious class of auditors, however, he recommended 'a full and scientific course of experimental philosophy on the plan generally adopted in universities'. The Managers concurred with this proposal, and Garnett was appointed Professor, with residence in the house, annual salary of £300, and prospect of increase to £500. Garnett also received the seal of approval from the landed interest shortly after this appointment by being elected an Honorary Member of the Board of Agriculture.[74]

The fashionable lectures were planned for twice a week in the afternoons, and the serious ones three times a week in the evenings. The programme began on 4 March 1800, in the room which is today the main library at the RI, since the Lecture Theatre was in the process of being built. Both series of lectures were popular, and the room was crowded throughout the season. Davy was hired as Garnett's lecture assistant on 16 February 1801, and the subject of the well-known cartoon by Gillray, 'New Discoveries in Pneumaticks!' (Plate 2) was taken from the afternoon lectures held that spring.[75]

Despite Garnett's popularity, the Managers refused him the raise he asked for in 1801. Bence Jones ascribes this to Garnett's error in publicly giving the French credit for Alessandro Volta's discovery of galvanism (obviously offensive during the war with Napoleon), and to his publication of the syllabi of his lectures without the Managers' consent. Both these actions may have caused friction between Garnett and the governors, but there is very little evidence as to what precisely led to Garnett's resignation. Davy was promoted to Lecturer in Chemistry on June 1st, and Garnett left two weeks later.[76]

The same summer saw the appointment of another Edinburgh M.D., Thomas Young, as Professor of Natural Philosophy, editor of the journal, and Superintendent of the House. Given the orientation of the RI, this was a peculiar choice, for Young was essentially a seventeenth-century scientist rather than a scientist of the Industrial Revolution. With a mastery of several languages by the age of seven or eight, Young was known to be eccentrically brilliant and completely independent. Though he was a physician by training, his

---

[73] The following is taken from *BJ*, pp. 167–70.

[74] RASE Archives, *Register of Official, Ordinary, Honorary, and Corresponding Members of the Board of Agriculture*, p. 44. The RASE was founded in 1838, and is still in existence. The materials (*Rough Minute Book* etc.) of the Board of Agriculture are preserved in its archives.

[75] Although a similar demonstration of the use of laughing gas is recorded as having taken place on 22 March 1800 by Lady Holland; *see The Journal of Elizabeth Lady Holland*, ed. the Earl of Ilchester, II, 60–1. The cartoon is discussed in detail by J. Z. Fullmer in *Scientific American*, CCIII (August 1960), 12–14.

[76] *BJ*, pp. 174–8. Garnett was undoubtedly angry, for he built a lecture room in Great Marlborough Street and set himself up as a rival to the entire RI. His courses included two on chemistry, three on experimental philosophy, two on botany (one of these at Brompton), and a course for medical students at Tom's Coffee House, in the City. He even edited a journal, *Annals of Natural History, Chemistry, Literature, Agriculture, and the Fine Arts*. Garnett died the following year.

true interests lay in pure physics, such as the wave theory of light, which he unsuccessfully attempted to resurrect during his tenure at the RI. His lectures there, published as *A Course of Lectures on Natural Philosophy and the Mechanical Arts* (1807), contain much on natural philosophy and little on the mechanical arts. In his very first lecture, 20 June 1802, Young stated that pure knowledge alone was enough, given the pleasure it provided, and that one need not ask for its practical applications.[77] Furthermore, he was not (unlike Garnett) interested in entertaining his audience, being, in the words of one colleague, 'worse calculated than any man I ever knew for the communication of knowledge'.[78] Thus he did not really fit in with any of the Institution's goals. The minutes of the Board of Agriculture reveal that at the same time that Davy was required to lecture to the Board on agricultural chemistry, the Managers of the RI offered

> to permit the Prof[essor] of Exper[imental] Phil[osophy] to give lect[ures] on ye mechan[ical] construc[tion] of ye diff[erent]t implem[ents] of Husb[andr]^y if it be deemed exped[ien]t by the B[oard].

The Board, furthermore

> considered^{ing} [*sic*] y^s [this] obj[ect] as [being] of very g[reat] imp[ortance] do thankfully accept ye offer of ye managers of the R[oyal] Inst[itution] & y^t [that] the Pres[iden]t of ye R[oyal] S[ociety] be requested to communicate ye same to them.[79]

Young must have refused to carry out this assignment, inasmuch as it did not take place; and it was not until a few years later that William Allen gave lectures on road transport and agricultural machinery at the RI in the capacity of Professor of Natural Philosophy. As in the case of Garnett, Young's resignation followed upon the refusal of the Managers to grant him an increase in salary, and he left on 4 July 1803 to resume his practice in medicine and devote his time to hieroglyphics. [80]

If the conflict with Garnett arose out of personal disagreements, and the one with Young out of ideological ones, then the choice of Davy proved to be ideal. It is not true, as has traditionally been maintained, that Davy's popularity as a lecturer saved the RI (it was bankrupt by 1803), or that he succeeded by being entertaining (so was Garnett). As I shall argue in Chapter 2, Davy was *malleable*. Like Garnett and Young, he had had some background in medicine (from Thomas Beddoes, at Bristol), but unlike them, he came to the RI at a very early age, and without any established career. The RI, at least until 1812, became his career, a development that would have been very difficult for Garnett or Young. Davy had his preferences, to be sure, but he was quite amenable to being shaped by circumstances, and his influence was derived

[77] *BJ*, p. 241.

[78] Quoted from a tutor at Emmanuel College, Cambridge, where Young was a fellow commoner, in *BJ*, p. 233.

[79] RASE Archives, *Rough Minute Book*, 1801–3, p. 134 (2 June 1802).

[80] Besides the biography by Peacock (*see* above, footnote 67), *see* Alexander Wood and Frank Oldham, *Thomas Young* (Cambridge University Press, 1954).

precisely from his identification with the goals of the governors. That influence became enormous as a result, whereas Garnett and Young found themselves out of their element very soon after their arrival.

To turn to the Lecture Theatre itself, construction began in April of 1800 and was completed in February of 1801, remaining unaltered until 1928.[81] When the house was purchased, it extended only part of the distance toward Grafton Street which it does now. Thus there was an empty space at the north end, and it was into this that the Theatre was expanded, on the second-floor level.[82] Thomas Webster was charged with designing it, although some controversy arose when the Managers also consulted an architect named Spiller. As Webster tells it, the circumstances that then arose were somewhat unpleasant. He resigned in the face of Spiller's alternative proposal (which would have required an expenditure of £10,000), was prevailed upon to stay, and saw his design finally adopted. George Saunders, architect of the British Museum, was also consulted, but in this case without conflict. The final contract came to £5,227 (for which £5,200 was raised from among the Proprietors), and involved building over the empty north end, and also demolishing this end of the building down to the second-floor level. On the north end also, and facing Albemarle Street, Webster built a stone staircase leading to a gallery within the Theatre, which he also constructed to accommodate a lower-class audience. By 1801, however, there was a significant change in attitude towards lower-class education. Webster records in his autobiography:

> ... I was asked rudely what I meant by instructing the lower classes in science. I was told likewise that it was resolved upon that the plan must be dropped as quietly as possible, it was thought to have a political tendency; if I persisted I would become a marked man.[83]

As a result, the staircase was eventually torn down.

In his autobiography, Webster claimed full credit for the design of the Theatre, although George Saunders apparently did have some say in its construction; and the Theatre does bear a great similarity to the one at Anderson's Institution, constructed in 1796. Whatever its inspiration, the auditorium proved to be an acoustic gem.[84] It also allowed the maximum number of persons clear vision for the given space, and was heated by steam for the convenience of the audience. Most important, its construction was something of a turning point in the history of science, for prior to February, 1801, there was no

---

[81] For the following see A. D. R. Caroe, *The House of the Royal Institution* (London: the Royal Institution, 1963); *BJ*, pp. 149ff.; and Thomas Webster's autobiography, preserved in bound MS at the RI, and written ca. 1837.

[82] i.e. the American second floor, British first floor. I shall use American terminology throughout.

[83] Quoted in Caroe, *House of the RI*, pp. 21–2, and *BJ*, p. 194.

[84] *See* Caroe, *House of the RI*, p. 25. Faraday also noted this in his testimony before a Select Committee of Parliament in 1835 (Report from the Select Committee on the ventilation of the Houses of Parliament. Ordered, by the House of Commons, to be Printed, 2 September 1835. *See also* the letter from Thomas Webster to Faraday dated 23 May 1836 in the Faraday Papers at the IEE.).

theatre in England that had been built expressly for the holding of scientific lectures.

The publication of a journal was also part of the attempt to diffuse practical scientific knowledge. Articles were not to be, like those in the *Philosophical Transactions of the Royal Society*, reports of private or antiquarian interest, but rather like the SBCP *Reports*, in the nature of information sheets, directly applicable to practical matters. Thus at the same meeting at which it was voted to have a journal (31 March 1799), the Managers appointed fourteen committees to investigate subjects such as soup, bread, cottages, stoves, utensils, fodder, fireplaces, furniture, machinery, and the like, with the results to be published in the journal.[85] Similarly, Davy abstracted a total of forty-seven articles from other journals, and most of these were on technical subjects.[86] Nevertheless, the journal was not able to escape the influence of the Royal Society tradition (as Brande recognized when he revived it in 1816), and did contain observations on comets, natural history, etc. Rumford supervised the first three issues, Young the next four, and was joined by Davy in 1802. The following year, probably owing to financial pressures, the journal was discontinued.

The Repository had even less success, for reasons that will be discussed in Chapter 3. Briefly, Matthew Boulton and other manufacturers saw the scheme as an attempt, on the part of the landed class, to benefit from their innovations free of charge, and became alarmed that the secrets of these inventions would be made public and thus ruin them financially. The Repository thus contained some of Rumford's stoves and fireplaces, as well as models of agricultural machinery, but industrial opposition was strong enough to render any further expansion impossible.[87]

Although the mechanics' school was not as stillborn as traditional accounts of the RI would have it, it is nevertheless the case that there was a reversal of feelings about the issue of lower-class education, already referred to, which made the failure of the school a foregone conclusion. Bernard and the SBCP, in the early days, viewed such education as a way of attenuating adverse political sentiment, and thus when Webster's proposal was read at a meeting of the Managers held on 14 September 1799, it was 'highly approved of'.[88] Even at this stage, however, there was more than one opinion on the subject, and Webster records that Banks was dubious about the wisdom of such a plan. Webster went to visit the latter to overcome 'a few political scruples which he had', and was apparently successful.[89] Webster himself was not worried, for he

[85] *BJ*, p. 153.

[86] J. Z. Fullmer, 'Humphry Davy's Critical Abstracts', *Chymia*, IX (1964), 97–115.

[87] *See* Ch. 3 below and *BJ*, pp. 184, 189, 196, and 204. Sparrow, *White Eagle*, discusses these events on pp. 127–30, and Davy referred to manufacturing-class opposition in his *Lecture, on the plan which it is proposed to adopt for improving the Royal Institution, and rendering it permanent* (London: William Savage, 1810), pp. 2–4.

[88] *BJ*, p. 143. See also *MM*, I, 58–60, as well as Sparrow's account, 'Early Days at the Royal Institution: A Forgotten Experiment in Technological Education', *Educational Review*, VI (1954), 202–7.

[89] *BJ*, pp. 144–5.

had given lectures to mechanics and was at that time in charge of a small school for artisans. 'I was not,' he writes in his autobiography, 'unacquainted with the political feelings of that time, but I did not think a little learning was a dangerous thing *if judiciously bestowed* . . ..'[90] His school at the RI embodied the principle that 'the men would be under the eye of the higher classes, and anything wrong would easily be put a stop to'.[91]

Despite conflicting currents, then, the school did get underway. A large room was set aside on the first floor, and bricklayers, joiners, and other tradesmen came and practised constructing chimneys, fireplaces and boilers, models of which found their way into the Repository, located on the first floor under the Lecture Theatre.[92] Lord Winchilsea, Thomas Bernard, the Palmerstons and others sent provincial tenants and cottagers to live at the RI to study with Webster, after which they 'returned to the part of the country from which they had come, and practiced what they had learned and taught others'.[93] Yet the scheme had been in trouble as early as September 1800. Writing to Garnett, Webster stated: 'I have not very little reason for supposing that such a plan will be at all put in practice . . .'.[94] The following year he wrote to a friend that sentiment for lower-class education had shifted considerably. 'I am not certain that the plan has been entirely abandoned,' he wrote, 'but I think that it is not likely to take place.'[95] The plan did, as indicated, take place, but it was eventually abandoned, and Webster left in 1802. The Managers did, nevertheless, retain the idea of lower-class education for a time. Artisans were still taught at the RI, first in the regular lectures and later in a series of evening lectures instituted by the Managers on 21 February 1803.[96]

These changes were not simple matters of disagreement with Webster or Rumford. It was during this period, as noted, that popular views on education for the poor began to change. For example, RI Proprietor William Wilberforce sponsored the Combination Acts, which made workers' organizations illegal and undoubtedly affected the climate in which the Institution developed. Davies Gilbert, friend of Davy and future Royal Society President, spoke strongly against the progressive view when Samuel Whitbread's bill on education was debated in Parliament. Gilbert

> was not alone in fearing that education would lead the poor to 'despise their lot in life, instead of making them good servants in agriculture and in the laborious employments to which their rank in society had destined them; instead of teaching them subordination, it would render them fractious and refractory . . . [it] would enable them to read seditious pamphlets, vicious books, and publications against Christianity; [and] would render them insolent to their superiors'.[97]

[90] ibid., p. 144.
[91] ibid.
[92] Caroe, *House of the RI, p. 21.*
[93] *BJ*, pp. 145–6.
[94] ibid., p. 163.
[95] ibid., p. 143.
[96] *MM*, III, 102–3.
[97] Quoted in Poynter, *Society and Pauperism*, pp. 215–16. Gilbert's name at this time was 'Giddy', a Cornish name which he later changed when he married in 1808.

By 1806 Patrick Colquhoun expressed the prevailing view when he wrote that 'science and learning, if universally diffused, would speedily overturn the best constituted government on earth'.[98]

The construction of the Laboratory, as of the Lecture Theatre, also constituted a major development in the organization of science, and a significant departure from the amateur tradition. We shall discuss its research function in Chapter 2, and how Davy gradually converted it into a place for agricultural analysis; but until 1803, its primary function was as an adjunct to the lectures. Research, however, was always recognized as a salient feature of its work. Thus in June of 1801 the Managers appointed a committee for chemical investigation and analysis that had Charles Hatchett as its chairman, and included Lord Dundas, Richard Chenevix, W. H. Pepys, Nicholson and Carlisle (famous for their experiment on the electrolysis of water in 1800), and two others. The following January, it resolved to undertake analyses on metallic alloys, although it is not clear if this work actually got underway.[99] There were also six workmen in the service of the RI, and a mathematical instrument maker, so that apparatus needed by professors for the lectures could be manufactured on the premises.[100] In April of 1803, however, Hatchett reported his dissatisfaction with the arrangement, complaining specifically that the Laboratory was too oriented to the role of service for the lectures.[101] He offered the following suggestions, adding that if implemented, the RI would boast a laboratory superior to any in England, and probably the Continent as well:

(1) Annexing the workshop to the Laboratory, and adapting the forge to chemical purposes;
(2) Building an air furnace and reverberatory furnace;
(3) Adding presses and shelves to contain vessels and additional material;
(4) Purchasing crude materials so that pure reagents could be manufactured instead of bought;
(5) Hiring a full-time assistant for Davy, and having the latter teach laboratory operations.[102]

As a result, a Committee of Science was appointed on 2 May 1803, consisting of Hatchett, the Earl Spencer, Sir Joseph Banks, Henry Cavendish, and one other governor, and by May 16th the Managers resolved to act on Hatchett's recommendations, including the construction of a lecture room next to the Laboratory, to accommodate 120 people.[103] This was completed by 1804, so that members could attend experiments; and Davy was also allowed to take on private pupils, as well as undertake analyses by commission, at £10 each.[104]

[98] ibid., p. 206.
[99] *BJ*, pp. 186 and 192.
[100] ibid., pp. 207–8.
[101] There is an 1803 inventory of apparatus in the RI Archives, Box File XIV, Folder 130, but there is no way of determining what was for use in the lectures as opposed to the Laboratory proper.
[102] *BJ*, pp. 214–15.
[103] ibid., pp. 215–18.
[104] ibid., pp. 258–9 and 269–71.

With these changes, the Laboratory did become outstanding; and by 1804, the RI had sponsored within a single building the leading laboratory in England, several courses of scientific lectures with live-in arrangements, a very large library, and (for a while) journals and a Model Room. Although such an undertaking required an occasional infusion of liberal loans and subscriptions, the RI did constitute a radical step in the direction of professional scientific organization.

To complete the picture of the very early years, it will be helpful to summarize the circumstances under which Rumford left, and the immediate aftermath. As already indicated, he implemented certain plans, but was not instrumental in their formulation or success; and, despite appearances, had a marginal role in the running of the RI. His final appearance at the meeting of the Managers on 26 April 1802 must have been in a very strained atmosphere. Rumford read a report that was more a description of what he believed he had done for the Institution than a statement of facts, and left for Paris a few days later.[105]

Attacks on Rumford began almost immediately and continued even after his death in 1815. The favourite word was 'charlatan'. In general, it was believed that his ideas on heat and fuel were either unsound or stolen from other sources, and that he had once again become something of a double agent politically, which included spreading the rumour among the French that Charles Blagden was a spy. There is no need to elaborate upon these attacks, as they do not concern the RI per se; but one issue is relevant to our story, viz. the private belief that Rumford had embezzled Institution funds. The first indication that something was amiss occurs in a letter Rumford wrote to Savage two weeks before the April 26th meeting. Apparently, one of the account books was lost, and Rumford's memorandum contains a note of urgency:

Get the Bankers to make out a new acc.' for the Committee of Expenditure *immediately today* if you should not yet have found the lost book.
If the book is found, carry it to the Bankers and get it written up. I must have this acc.' tomorrow . . . .
Everything must be quite ready for Monday—I must see you today.[106]

One month later, Rumford wrote to Joseph Banks that he had left with him all the bills to be paid. Banks made an itemized list of these bills under the heading, 'bills Left with me by Count Rumford' on the letter itself, and the total came to more than £1,900.[107] Bence Jones records that in April 1801, Treasurer John Hippisley reported a surplus of £10,000 to £11,000, whereas one year later, when Rumford departed, there was a deficit of £720.[108]

The Managers ultimately instructed Savage to write to Rumford about the matter, and the Count replied from Paris in 1803.

[105] ibid., pp. 195–7.

[106] RI Archives, Rumford MSS, Letter 5 (10 April 1802); italics Rumford's.

[107] ibid., Letter 6 (7 May 1802). Banks did not sign his name, but his handwriting is easily identifiable.

[108] *BJ*, pp. 180 and 200.

I assure you that I have not the smallest recollection of having received from Mr Hunter [solicitor to the RI] the Account you mention in your letter of the 7th Ult[imo]; and had I in fact received it I should most undoubtedly have laid it before the managers of the Institution. I can imagine no reason which could have induced me to keep it back; and as all the affairs of the Institution in my hands were kept with the utmost care and regularity, as you can testify, it is not likely that I should have mislaid and forgotten it. This is all I can say on the subject, and I hope and trust that this declaration will be satisfactory to the Managers of the Royal Institution, and to Mr Hunter.[109]

The following year, replying to a letter which Savage wrote to him on 7 June 1804, Rumford told the latter: 'With regard to the outstanding bills which you mention, as I intend and expect to come to England soon they may as well stand on till my arrival'.[110] But he never did return.

These discrepancies are even more suggestive in the context of several entries in Blagden's diary. In June of 1802, one month after Rumford had left England, Blagden met one of the RI Managers, Alexander Blair, in Paris. Blair told him

That C^t Rumford had been violent at [the] meeting of proprietors of [the] R.I. because they had wanted to look into the accounts; [Rumford] said they had nothing to do with them.[111]

A year later, back in London, Blagden visited Banks, who, he wrote, 'told me how the accounts of the RI had fallen short of R[umford]'s statement ...'.[112] Shortly after that, Blagden was at Lady Palmerston's where he was joined by Sir Henry Englefield.

[I] mentioned to her C[oun]t R[umford]'s letter, asked if she knew any thing of his supposed marriage afoot: she said no; desired her not to mention it to him. Sir H. Englefield's story of applications for money, put [?] to keep out of jail; then to get married, & the advice against his wife's [Madame Lavoisier's] bad temper, which found to be only a trial of cash.[113]

The implication here, that Rumford had to come up with funds to avoid imprisonment, is perhaps not surprising. In Madame Lavoisier he once again found a wealthy widow, and the marriage was commonly regarded as a lucrative venture for the Count. Such evidence is, however, at best circumstantial, and much more to the point is an entry in Blagden's diary of about one year later, referring to Sir Francis Baring. Baring, who was Chairman of the East India Company and a leading London banker, subsequently led a breakaway movement from the RI which became known as the London Institution; and the following set of events may clarify, in part, his motivation. Blagden visited Banks on 3 June 1804, and found Henry Brougham and Robert

[109] Rumford MSS, Letter 12 (11 November 1803).
[110] ibid., Letter 13 (3 July 1804).
[111] Blagden, *Diary*, 5 June 1802.
[112] ibid., 9 May 1803.
[113] ibid., 28 May 1803.

Livingston, the American Ambassador to France, present at the Soho Square apartments. 'Brougham,' wrote Blagden, 'gave strong intimation, that C[oun]t R[umford] stole [his ideas] from [Sir John] Leslie,' and recorded that Banks agreed. Apparently, this loosened Livingston's tongue. On the way home, he confided to Blagden that 'Baring had told him that C[oun]t R[umford] had embezzled some of the funds of the Institution'.[114] Bence Jones did not mention the matter in his history of the RI, but wrote to George Ellis that he was 'much mistaken if the managers did not suspect the accounts "had been cooked", so to say, for they called in an accountant'.[115] As it turned out, the issue never got much beyond the RI itself, either because absolute proof was lacking, or because of the desire on the part of the governors—some of the greatest names in Britain—to avoid a scandal. Embezzlement and silence constituted a common pattern among eighteenth- and nineteenth-century institutions, and the RI was, moreover, at a crucial juncture, staving off bankruptcy by a £2,000 loan from the governors in 1803. Negative publicity could have ruined the Institution at this time, and thus it was desirable to confine knowledge of the embezzlement, if it indeed took place, to the inner circle at the RI as best as possible.

In summation, then, the foundation of the RI, and its early history, are not to be attributed to the activities of Count Rumford. The dual revolution created new attitudes to science, agriculture, class tensions, and education; and institutions like the Board of Agriculture, the SBCP, and the RI were the visible manifestations of these attitudes. Even within the RI, specific developments that made the Institution unique were part of the development of resources to deal with new changes and social problems. Rural poverty, for example, is not just an 'interesting backdrop' to the RI; instead, it generated the need for a different kind of institution and influenced the nature of institutional activity. It is not that personalities did not matter in the day-to-day activities of the Institution, but rather that the RI derived its essential character from factors far more pervasive than the desires of any single individual. The critical factor for the first decade proved to be the 'revolution' in British agriculture and the class interest that underlay it. It is to this issue that we must now turn.

---

[114] ibid., 3 June 1804.
[115] Ellis, *Rumford*, p. 440.

2
The 'Society of Husbandry ... in Albemarle Street'

If one side of the dual revolution created, for some members of the upper class, an interest in scientific philanthropy, the other side created a similar interest in the development of agriculture. The first issue, in the form of the SBCP, brought the Royal Institution into being; the second, in the form of the Board of Agriculture, provided it with a direction in which to grow. As soon as philanthropy became less of a preoccupation (ca. 1801), the RI began, as a matter of course, to get involved in issues that were more immediately relevant to the landowning class. Specifically, the failure of allotment, the change in opinion regarding education for the poor, and the inability of the Board to cheapen the process of enclosure[1] gradually led the RI away from the philanthropic aspects of agricultural development to the more remunerative ones. To understand the subsequent history of the RI, then, it will be necessary to set the Institution aside for the moment and examine the economic picture of the time as it relates to the question of agricultural production.

By 1793, when the Board of Agriculture was created, the lucrative possibilities of improved husbandry were optimal. They had, of course, been building for some time. The late eighteenth century saw crucial improvements in agriculture and the massive enclosure of land by Acts of Parliament which affected six million acres.[2] Although the role of the landowners in all this may have been exaggerated, some did interest themselves in various facets of estate activity. Poor harvests, growing industrialization, and rising national population resulted in increased demands for food; and by augmenting the yield of his lands, the proprietor hoped to increase his rent rolls. Between 1750 and 1790 40% to 50% increases in rents were typical of many large estates, and these higher incomes were in turn an incentive for further

---

[1] The General Enclosure Act of 1801 was a victory only on paper, and its provisions easily emasculated by the Church and the Bar, which were concerned about tithes and fees. See J. D. Chambers and G. E. Mingay, *The Agricultural Revolution 1750–1880* (London: B. T. Batsford, 1966), p. 121.

[2] See Ch. 1, footnote 3, and Chambers and Mingay, *Agricultural Revolution*, p. 77. About one quarter of the cultivated acreage was affected during the reign of George III. Enclosure, however, had begun in the fifteenth century and was normally done by agreement. The formal use of Acts of Parliament on a large scale began after roughly half of the agricultural land of England was already enclosed (ca. 1700).

5a. Arthur Young, 1794

5b. Earl Spencer

6. 'The Generae of Patriotism,' by Gillray

agricultural progress.[3] Some of this had a fashionable aspect to it. George III enjoyed being called a farmer and published (under the pseudonym of Ralph Robinson) on agricultural subjects in Arthur Young's *Annals of Agriculture*. Sheep shearings and agricultural exhibitions were social occasions as well as opportunities for learning about new improvements. The interest in improvement which 'began to be conspicuous in the upper ranges of the social scale' thus became a vehicle for even greater status and prestige.[4]

With the outbreak of war in 1793, the embargo on grain importation, and the unusually poor harvests of the 1790s, the price of grain rose and these trends were intensified.[5] The Napoleonic era marked the most intense period of enclosure by Act of Parliament in English history—2,000 Acts between 1793 and 1815.[6] By 1793, according to Asa Briggs,

> not only fashion but the hope of economic gain was impelling the landlords to interest themselves in agricultural improvement ... business-like domestic farming became a reasonable economic proposition .... The sun shone for the landlord during this time far more brightly than it had done for many years, and he hastened usually not to make hay but to grow corn.[7]

James Gillray captured the events of the decade in his cartoon 'The Generae of Patriotism', depicting the Duke of Bedford sowing gold coins into the soil (Plate 6); and the following excerpt from a letter from the Earl of Egremont to Arthur Young testifies to the change among some of the upper class: 'Tempted by the high prices and the exemption from the Tythe [*sic*] which I have purchased I am going to convert some more of the fine soil of my park into arable [land] . . .'.[8]

Nevertheless, this group was not typical of the landed interest. It was with obvious justification that leading agricultural writers constantly complained of the indifference of the great proprietors to the new husbandry. The majority of estates were cultivated quite conventionally, as units of management rather than production, with the management largely in the hands of stewards and land agents. Landowners were conservative, and as the bulk of their income was derived from rents, they tended to be more interested in securing efficient

---

[3] Peter Mathias, *The First Industrial Nation* (London: Methuen, 1969), pp. 57 and 73, and G. E. Mingay, 'The Large Estate in Eighteenth-Century England', First International Conference on Economic History, Stockholm, 1960, *Contributions* (Paris: Mouton, 1960), p. 377. *See also* Phyllis Deane, *The First Industrial Revolution* (Cambridge University Press, 1965), pp. 42 and 48.

[4] Deane, *First Industrial Revolution*, p. 46.

[5] Mingay, 'The Large Estate', p. 377. All the harvests from 1795 to 1800 were poor, notably those of 1795 and 1799. The price of wheat rose from 43s. per quarter (British imperial quarter = eight bushels) in 1792 to 108s. 4d. by August 1795 to 119s. 6d. in 1801. Corn was 55s. 6d. in 1791, 142s. 10d. in 1800. *See* Chambers and Mingay, *Agricultural Revolution*, pp. 112–14; W. E. Minchinton, 'Agricultural Returns and the Government during the Napoleonic Wars', *The Agricultural History Review*, I (1953), 29 and 37; and Paul Mantoux, *The Industrial Revolution in the Eighteenth Century*, trans. Marjorie Vernon (rev. ed.; New York: Harper and Row, 1961; original French ed. publ. 1928), p. 347n.

[6] Chambers and Mingay, *Agricultural Revolution*, p. 77.

[7] Asa Briggs, *The Making of Modern England* (New York: Harper and Row, 1965), p. 40.

[8] BM Add. MS. 35128, f. 202. The letter is dated 10 March 1800.

7. The Duke of Bridgewater and his canal

tenants than in experimenting with innovative techniques.[9] The Board of Agriculture, which contained most of the really dedicated landlords and farmers, failed in its attempt to get the rest of the upper classes interested in scientific husbandry.[10] Yet for those dedicated few, the economic opportunities of the age, and the technical improvements that were available, amounted to a significant difference in their way of life.

For those interested, farming was not the only path to profit or increased rents, for agricultural changes were part of the larger process of industrialization. There was, in fact, no sharp dichotomy between agricultural and non-agricultural activity at this time. The investment in minerals under one's land, or in transport facilities such as canals and roads, was a logical extension and legitimate part of estate activity. Landowners and their stewards 'did not think of land in a purely agricultural sense', but became involved in the exploitation of mineral deposits, agricultural products (e.g. hides), and forests, which served as a source of wood for furniture or of bark for tanning leather.[11] The 1790s have been labelled a period of 'canal mania', and it was 'natural ... for the landowner to join with the merchant and industrialist in projects to cheapen and speed up the means of transport ... '.[12] As in the case of farming, industrially oriented landowners were the exception rather than the rule. Professor Habakkuk has suggested that the Duke of Bridgewater, traditional founder of inland navigation in England, might have been 'a freak'.[13] Yet while recent research has tended to indicate that the part played by the great proprietors in agricultural developments has been exaggerated, it also suggests that their importance to industrial activity has been overlooked.[14] As the Industrial Revolution gained momentum, entrepreneurial landlords tended to become rentiers; but in the late eighteenth century some landowners were 'still among the leading entrepreneurs'.[15] Most of the estate exploitation was handled by stewards and land agents, yet the

maintenance of these conventions was often mingled incongruously with a sharp

[9] Mathias, *First Industrial Nation*, pp. 57–8; G. E. Mingay, *English Landed Society in the Eighteenth Century* (London: Routledge and Kegan Paul, 1963), pp. 167–8.
[10] Ernest Clarke, 'The Board of Agriculture, 1793–1822', *The Journal of the RASE*, Third Series, IX (1898), 1–41, and Rosalind Mitchison, 'The Old Board of Agriculture (1793–1822)', *English Historical Review*, LXXIV (1959), 41–69. Mitchison also discusses the Board in *Agricultural Sir John* (London: Geoffrey Bles, 1962), as does John Gazley in *The Life of Arthur Young* (Philadelphia: American Philosophical Society, 1973), pp. 306–59. *See also* the dissertation by Winifred Harrison cited in Ch.1, footnote 14, above.
[11] Mingay, *English Landed Society*, pp. 190–1.
[12] ibid., p. 196.
[13] H. J. Habbakuk, 'Economic Functions of English Landowners in the Seventeenth and Eighteenth Centuries', *Explorations in Entrepreneurial History*, VI (1953), 97.
[14] Mingay, 'The Large Estate', p. 383. In his recent study of canal financing, J. R. Ward found that nearly one fourth of the canals in his sample, built during the period 1755–1815, were financed by landowners and farmers (*The Finance of Canal Building in Eighteenth-Century England* [London: Oxford University Press, 1974]).
[15] Mingay, *English Landed Society*, p. 201. *See also* F. M. L. Thompson, 'English Great Estates in the 19th Century, 1790–1914', First International Conference of Economic History, Stockholm, 1960, *Contributions* (Paris: Mouton, 1960), p. 396.

awareness of the economic opportunities of the age. Some exceptionally energetic members of the aristocracy entered into an enormous range of projects which went far beyond the bounds of estate exploitation.[16]

The fact that interests were generally handled for the landlords by their stewards

> was all the more reason to supervise their affairs closely, and many of them made the decisions that are the hallmark of active entrepreneurship. A significant few opened mines, built iron works and mills, dug canals, developed ports, and leased their urban properties for building.[17]

As Mingay put it, ' some of the more ambitious landowners were willing to consider any project which offered prospect of profit'.[18]

In industrial as well as agricultural activity, it is impossible to get more than a qualitative impression of the size of this group. As to its effectiveness, it is unlikely that the improvers were the historical agents of an 'agricultural revolution' in the late eighteenth century, and as time passed the level of their industrial involvement tapered off. But what distinguished this group as atypical was not merely its activities, but its attitude. By all accounts, its impact was out of all proportion to its size. The emergence of such a group, especially when many of its members were so wealthy that agricultural and industrial profits were marginal to their economic interests, bears witness to the powerful psychological changes that had occurred in response to the early Industrial Revolution. The significance of their efforts, writes David Landes, 'lay in the efforts themselves, not in their return ... it lay in the legitimacy conferred on innovation and the pursuit of wealth as a way of life'.[19] It is this assessment, and that of G. E. Mingay, that most accurately captures the role and psychology of the improving landlords:

> [I]t was perhaps their *attitude* towards economic development that was of greater significance than their own direct efforts ... they fanned the flame of enterprise and innovation and were not merely passive spectators of the age of change.[20]

Concomitant with this attitude and activity was a technological conception of science, a Baconian and entrepreneurial ideology that represented an important alternative (not yet a rival) to the gentleman amateur tradition. It was almost inevitable that an industrial redefinition of scientific activity and its purposes would emerge in the late eighteenth century; but perhaps less expected was that this ideology would manage to affect a part of the aristocracy. This upper-class interest in science as a utilitarian activity is surprising in that it did not properly 'belong' to the aristocracy; but then neither did bourgeois at-

[16] Mingay, 'The Large Estate', p. 381.
[17] David Landes, *The Unbound Prometheus* (Cambridge University Press, 1969), p. 69.
[18] *English Landed Society*, p. 190.
[19] Landes, *Unbound Prometheus*, p. 69.
[20] *English Landed Society*, p. 201; italics in original.

titudes of entrepreneurship. We are not, however, tied to any axiomatic formulation of ideology: Marx's famous statement that the 'class which has the means of material production at its disposal, has control at the same time over the means of mental production' is surely too facile.[21] Michel Foucault's comment on this issue is far more astute. Although 'membership of a social group can always explain why such and such a person chose one system of thought rather than another,' he writes, 'the condition enabling that system to be thought never resides in the existence of the group'.[22] For most of the aristocracy science quite naturally meant the gentleman amateur tradition. One was, for example, invited to the stiff, dignified weekly meetings at Banks' house on Soho Square to admire some shells or curiosities.[23] The question for us, then, is what enabled a part of the upper class to develop an entrepreneurial ideology of science? What was 'the condition enabling that system to be thought'?

The answer to this becomes apparent as soon as we leave London and move to the northern provinces where, with the advent of the Industrial Revolution, the Baconian definition of science had experienced a strong revival. For the new class of technicians and industrial magnates, 'the arts'—innovations, inventions, and other technological improvements—were dependent upon science and no real distinction was made between scientific and technical knowledge. 'Science' now meant the fruit of inventiveness, a combination of technology and applied science. It was no longer conceived of in the context of polite learning or disinterested research, but rather as closely linked to the production of wealth. Whether or not scientific theory and experiment could actually contribute to this rising industrial capitalism is much beside the point, for it was generally accepted that the two were intimately bound up, and it was on this perception that men began to base their actions. This, among other reasons, accounts for the heavy diffusion of technology during the Industrial Revolution. The impressive list of clubs, societies, itinerant lecturers, and scientific publications compiled by A. E. Musson and Eric Robinson bears witness to the intense interest in science and technology sweeping the provinces during the late eighteenth century. The idea of a marriage between science and industry envisaged by Bacon, which had been almost totally submerged in the course of the eighteenth century, was now making a dramatic comeback.[24]

The affinity, then, of the entrepreneurial segment of the aristocracy with this ideology of science is not difficult to understand. For the group we have been discussing science began to mean more than the reportage of curiosities in the *Philosophical Transactions*. It represented not just wealth, but a reliable way

---

[21] Karl Marx and Frederick Engels, *The German Ideology*, ed. C. J. Arthur (London: Lawrence and Wishart, 1970), p. 64.

[22] Michel Foucault, *The Order of Things* (New York: Vintage Books, 1973), p. 200.

[23] Perhaps a misleading example, since Banks was a complex figure who embodied both the amateur tradition and the view of science as a key to profitable estate activity.

[24] A. E. Musson and Eric Robinson, *Science and Technology in the Industrial Revolution* (Toronto: University Press, 1969). In my review of their work, I tended to see the marriage-of-science-to-industry theme as something of a mass delusion (*Journal of Social History*, V [1972], 521–7).

of obtaining it. Thus science was seen as the rational understanding and methodology that underlay the steam engine of Boulton and Watt, the bridges and mills of John Rennie, the construction of canals and turnpikes, the bleaching of cloth and the glazing of pottery. In agriculture it meant experimenting with different crops and manures, inventing new farm implements, developing a better breed of sheep. As a result of this redefinition, and the energetic activity of a number of agricultural improvers, the 'relevance of the scientific method to the extension of agricultural knowledge was widely recognized by the end of the eighteenth century'.[25] In the area of agriculture in particular, this attitude succeeded in crystallizing in a specific organization, viz. the Board of Agriculture, 'the climax of the late eighteenth-century improvers'.[26] The first President of the Board, Sir John Sinclair, collaborated with Joseph Banks on sheep-breeding experiments,[27] and the second President, Lord Somerville, set aside twenty-eight acres of land for chemical research on soil analysis and food production. Lord Somerville was insistent that the Board, to succeed, must be supported by men of scientific interest, men 'who have grafted theory on approved practice'.[28] Arthur Young summarized the Board's objectives (as stated by Sinclair) when he noted that the Board could not tell the landowners and farmers what to do, but could 'encourage a spirit of experiment ... '.[29] 'Science is never better or more liberally employed,' he wrote, 'than when exerted in pursuits that assist the more necessary occupations of mankind.'[30]

The fortunes of the improving landlords were hardly dependent on science as such, yet they did see it as relevant to their interests for a number of complex reasons: the association of science with new sources of wealth (such as the steam engine); the desire to share in a fashionable interest in the subject; the belief that innovation and improvement contributed to the national good and were thus patriotic; and perhaps a vague fear that another social class was appropriating the subject to its own special advantage. It is thus more correct to speak of an *ideology* of science rather than merely an attitude towards it, for science was becoming part of a whole new value system and way of life.

It would be incorrect, of course, to believe that all the members of the Board were uniformly active in agricultural improvement or other estate activities. There was, rather, a spectrum of intensity in such matters, ranging from men

---

[25] Harrison, *The Board of Agriculture*, pp. 37–8.

[26] G. E. Fussell, 'Science and Practice in Eighteenth-Century British Agriculture', *Agricultural History*, XLIII (1969), 15.

[27] Mitchison, *Agricultural Sir John*, pp. 112–18, and H. B. Carter, *His Majesty's Spanish Flock* (Sydney: Angus and Robertson, 1964). On Sinclair *see* Ch. 1, footnote 18 above, and also James E. Handley, 'Sir John Sinclair', *The Innes Review*, VIII (1957), 5–18, and A. and N. L. Clow, *The Chemical Revolution* (London: The Batchworth Press, 1952), pp. 362 and 364.

[28] John Somerville, *The System Followed During the Last Two Years by the Board of Agriculture* (2nd ed.; London: W. Miller, 1800), p. 4.

[29] *AY*, XXI (1793), 130.

[30] ibid., pp. 230–1. The definitive work on Young is the biography by John Gazley (above, footnote 10). *See also* Ernest Clarke's sketch in *The Journal of the RASE*, Second Series, LIX (1893), 1–17; Henry Higgs, 'Arthur Young', *DNB*, LXIII (1900), 357–63; and M. Betham-Edwards (ed.), *The Autobiography of Arthur Young* (London: Smith, Elder, 1898).

like the Duke of Bridgewater, who were genuine entrepreneurs, to those whose interest was undoubtedly more casual. But the Board did make a serious effort to maintain a high level of participation on the part of its members. The thirty-one Ordinary Members did for the most part represent a strong commitment to agricultural improvement, and to insure this the Board annually dropped the five least active and replaced them with five of the Honorary Members, who 'were chosen from the more enthusiastic gentry and farmers ... '.[31] The 'Honorary' title should not mislead us, as these men were often as active or more active than the Ordinary Members. They paid a subscription fee of ten guineas, attended meetings, and after 1800 were permitted to participate in debates.[32]

The relevance of the Board of Agriculture for the Royal Institution becomes clear upon examination of the Proprietorship of the RI. If we consider the first fifty-seven Proprietors, for example, we find that eleven were Ordinary or Official Members of the Board of Agriculture, and fifteen were Honorary Members—i.e., more than 44% of the original Proprietors were, in varying degrees, atypical in their interest or actual participation in estate activity.[33] The spectrum of activity becomes most intense when we examine the first elected Managers and Visitors of the RI. Of the nineteen governors, fourteen belonged to the Board of Agriculture and eight (discussed below) were among the most outstanding agricultural improvers or industrial entrepreneurs of the age. Nor did the situation change much during the first decade. In effect, the two institutions had the relationship of an interlocking directorate. The interests, values, and ideology of science of the RI essentially emerged out of the industrial and agricultural changes of late eighteenth-century England.

The extent to which this was so becomes evident upon a close analysis of the improving landlords among the governors. Sir Joseph Banks, for example, was (despite his Presidency of the Royal Society) very representative of the new conception of science.[34] Agricultural improvement was a major part of his life. He worked with Sinclair on sheep breeding, wrote on the problem of increasing

[31] Mitchison, 'The Old Board of Agriculture', p. 43.

[32] ibid. There was an arbitrary number of Honorary Members (eventually 520). Besides the Honorary and Ordinary Members, there were also sixteen Official Members, e.g. the Archbishop of Canterbury, the Secretaries of State, etc. who for the most part did not represent any real interest in agricultural improvement. A few, however, were notable exceptions, such as Sir Joseph Banks and the Bishop of Durham.

[33] The Ordinary and Official Members are Joseph Banks, the Bishop of Durham, the Earl of Egremont, the Earl of Mansfield, Thomas Pelham, W. M. Pitt, John Sinclair, Lord Somerville, the Earl Spencer, William Wilberforce, and the Earl of Winchilsea. The Honorary Members are Thomas Bernard, Rowland Burdon, Joseph Grote, R. B. Harvey, J. C. Hippisley, Henry Hoare, Lord Hobart, William Lushington, John Macpherson, the Earl Morton, James Pulteney, J. B. Riddell, G. L. Staunton, R. J. Sulivan, and Samuel Thornton. One Proprietor who was an improving landlord but not on the Board was Lord Keith. The total percentage of improving landlords among the first fifty-seven Proprietors is 49.1%.

[34] On the following *see* above, footnote 27; *Public Characters*, III (1801), 370–401; B. D. Jackson, 'Joseph Banks', *DNB*, III (1885), 132; Elie Halévy, *History of the English People in the Nineteenth Century*, Vol. I: *England in 1815*, trans. E. I. Watkin and D. A. Barker (2nd ed.; London: Ernest Benn, 1961 reprint), p. 228; Mingay, *English Landed Society*, pp. 173–4; and Gazley, *Life of Arthur Young*, pp. 215–17, 379–80, and 491.

wool production, and organized the importation of Merino sheep for the nobility and the King. He ran his Revesby estate like an expert administrator, and managed to derive an annual income of £6,000 from it. Banks was also interested in industrial ventures, for he drew up a lease with the Duke of Devonshire to mine the lead vein on the latter's estates, and ordered machinery from the Boulton and Watt Company to assist him in this effort.[35] Banks' influence on the Royal Institution during the early years was greater than that of any other individual. He arranged for Davy to lecture the Board of Agriculture on agricultural chemistry and guided his studies on tanning.

Peter Mathias identifies two RI governors, the Bishop of Durham and the Duke of Bridgewater, as leading industrial entrepreneurs. [36] The Bishops of Durham had great temporal powers in Yorkshire until 1836, and were active in the development of coal mining. Sir Timothy Eden wrote of our own bishop, Shute Barrington, who was President of the SBCP: 'thanks to the steady development of coal throughout the eighteenth century, [he] was, by this time (1791), a very wealthy man'.[37] As for the Duke of Bridgewater he was, by all accounts, the epitome of the small group of innovators and improvers we are discussing. As 'the first entrepreneur of the canal age', his personal commitment and resourcefulness were vital to the construction of his canal.[38] The Duke owned land at Worsley, near Manchester, in which there were significant coal deposits. By having his engineer, James Brindley, cut a canal from Worsley to Manchester, he caused the price of coal at Manchester to drop 50%. To his death he remained the leader of the canal construction movement, spending £220,000 on his canals and eventually realizing an annual income of £80,000 as a result. Part of his estate was inherited by his nephew, the Earl of Bridgewater, an agricultural improver who continued building canals for the transport of produce as well as coal and ore.[39] It is significant that the Managers asked the Duke to become President of the Royal Institution, for he

[35] The lease is in BM Add. MS. 6679, ff. 5–10, dated 1779. The mines were known as the Gregory Mines. Information on the machinery ordered is in the Birmingham Public Libraries' Boulton and Watt Collection, *Catalogue of Old Engines*, pp. 284–5. It is recorded that the drawings for the engine for 'Messrs. Banks Milnes & Co' were begun in 1779, with a further comment: '*Gregory Lead Mine*, in the parish of Ashover, Derby. Joseph Banks of Soho Square Westminster William Miles & Samuel Kirk, both of Ashover & other adventurers'. On pp. 56–7 it is recorded that a colliery winding engine was made for the Duke of Devonshire in 1788. Permission to quote from this source is by courtesy of the Birmingham Public Libraries.

[36] Mathias, *First Industrial Nation*, p. 156.

[37] Timothy Eden, *Durham* (2 vols.; London: Robert Hale, 1952), II, 381–2. *See also* G. D. H. Cole and Raymond Postgate, *The Common People 1746–1946* (4th ed.; London: Methuen, 1968), pp. 17 and 21.

[38] H. J. Dyos and D. H. Aldcroft, *British Transport* (Leicester: University Press, 1969), p. 87.

[39] Francis Espinasse, 'Francis Egerton', *DNB*, XVII (1889), 151–3; G. E. Cokayne, *The Complete Peerage of England, Scotland, Ireland, Great Britain, and the United Kingdom*, ed. Vicary Gibbs (new ed., rev. and enl.; London: The St. Catherine Press, 1912), II, 314; Mantoux, *Industrial Revolution*, pp. 123–5; Witt Bowden, *Industrial Society in England towards the End of the Eighteenth Century* (2nd ed.; London: Frank Cass, 1965), pp. 63–4; and W. H. Chaloner, 'Francis Egerton, Third Duke of Bridgewater (1736–1803): A Bibliographical Note', *Explorations in Entrepreneurial History*, V (1952), 181–5. On the Earl of Bridgewater *see* Cokayne's *Peerage*, II, 315, and the *Annual Register*, XLVII (1805), 430.

probably represented an ideal to which they themselves aspired.[40]

The Earl of Egremont has long been recognized as one of England's great improving landlords. The Managers chose him to be one of their Vice-Presidents. Egremont

> converted eight hundred acres of Petworth Park into a model farm, and his methods in drainage and stock-breeding, his cattle show, and his use of new tools and turnip cultivation earned the praise of no less a critic than William Marshall.[41]

The Earl was convinced of the value of science for agriculture and used 'Mr Davies mixture to accelerate vegetation' for his turnip seeds.[42] Egremont promoted the Rother and London-Portsmouth canals, and 'was a chief mover in the Arun navigation, the Wey and Arun Junction, and the Portsmouth and Arundel canals'.[43] Despite his fabulous wealth—he was reputed to have given £1 million to charity during his lifetime—Egremont was fully aware of the economic possibilities around him and quick to take advantage of them.

We have already mentioned the interest of Lord Somerville in agricultural chemistry and his belief in the importance of science for farming. He was, in fact, one of the most active landlords in agricultural improvement.[44] He helped found the Smithfield Club for the improvement of stock, and in 1802 began to sponsor annual cattle shows, with competitions and prizes. He converted his Somerset estate into a profitable farm and realized an 11% return on his investment. Lord Somerville also invented or improved several farming implements and devoted himself to the improvement and breeding of sheep.[45]

For the Spencers, estate activity became a family tradition. The first Earl supported the Coventry and Oxford canal project in 1768.[46] The second Earl, with whom this study is concerned, became a leading sponsor of the Grand Junction Canal project during 1793–1805, which provided the main link from London to the industrial Midlands.[47] An Official Member of the Board of Agriculture, he had Joseph Banks supply him with Merino sheep for breeding.[48] Le Marchant wrote of him:

---

[40] *MM*, I, 30 (27 April 1799) and 32 (4 May 1799). He declined and became a Visitor instead.

[41] Mingay, *English Landed Society*, p. 164. *See also* H. A. Wyndham, *A Family History, 1688–1837*, and G. Le Grys Norgate, 'George O'Brien Wyndham', *DNB*, LXIII (1900), 224–6. Egremont was one of Arthur Young's closest friends (*see* Gazley, *Life of Arthur Young, passim*).

[42] With ill success: *see* below, pp. 59–60. Young records many of Egremont's experiments in *AY*.

[43] Mingay, *English Landed Society*, p. 199.

[44] ibid., p. 164.

[45] Ernest Clarke, 'John, Fifteenth Lord Somerville', *The Journal of the RASE*, Third Series, VIII (1897), 1–20; 'John Southey Somerville,' *DNB*, LIII (1898), 253; *Public Characters*, IX (1807), 202–3, 206, 212, and 221–2. Somerville's inventions and improvements included the double-furrow plow, a friction-drag to prevent too rapid descent of wagons on inclines, and a new colter. See *AY*, XXXIII (1799), 76–8, XLII (1804), 72–7, and XLIII (1805), 641–52; and *Communications to the Board of Agriculture*, II (1800), 415–23.

[46] Mingay, *English Landed Society*, p. 198.

[47] Mathias, *First Industrial Nation*, pp. 111 and 117–18.

[48] *See* the letters to the Earl from Lord Somerville (3 June 1806) and Joseph Banks (e.g. 31 July 1809) at the Muniment Room at Althorp, Northampton.

Lord Spencer entered heartily into the movement made at the beginning of the century, for the more scientific treatment of agriculture ... the example of the farm at Althorp (no insignificant item in his expenditure), where many experiments were made and new methods adopted, was not without good effect in Northamptonshire.[49]

His son, the third Earl, founded the Royal Agricultural Society in 1838.

Rowland Burdon, a wealthy merchant, banker, entrepreneur and Honorary Member of the Board of Agriculture, is best known for building the largest cast iron bridge in England across the River Wear (in the heart of the coal district) between 1793 and 1796.[50] He obtained the idea of using iron in bridge construction from similar attempts made by the Coalbrookdale Company, and supplied £30,000 for the project. Since the span of the river was so great, a new method had to be devised for building an arch. Architects doubted that it could be constructed, which it would not have been 'if the Inventive Talents of Mr. Burdon had not been equal to his Public Spirit'.[51] The resulting bridge was longer than the one built by Abraham Darby at Coalbrookdale. In 1802 Burdon and his engineer, Thomas Wilson, patented the methods they employed for combining and connecting the iron blocks used in constructing the arch.[52] Burdon himself made an experiment with one rib of the arch previous to undertaking full construction. The bridge was only part of a general plan he had for 'forming a complete Inland Communication' through Durham, linking the commercial towns, and he also invested in the building of turnpike roads. As he possessed a large estate in the area, he had much to gain from the establishment of a new transport network.[53]

The President of the Royal Institution, the Earl of Winchilsea, an Ordinary Member of the Board of Agriculture, was also famous as an improving landlord. His time 'was in great measure devoted to agricultural pursuits'.[54] He was the only landowner in Rutlandshire to use the drill (a type of shallow furrow) in sowing turnips, and was, as a result, known for his fine crops.[55] He also carried out experiments on the early sowing of oats, and on the feeding of sheep in an attempt to increase their weight.[56] Arthur Young was so impressed

[49] Denis Le Marchant, *Memoir of John Charles Viscount Althorp Third Earl Spencer* (London: Richard Bentley and Son, 1876), pp. 13–14.

[50] Joshua Wilson, *A Biographical Index to the Present House of Commons* (London: Richard Phillips, 1806), pp. 99–100; *Annales des Arts et Manufactures*, II, 166–73; *Annual Register*, LXXX (1838), 224–55; Thomas Bowdler, 'An Account of an Iron Bridge lately erected near Sunderland, by Rowland Burdon M.P. In a Letter from Tho: Bowdler Esq'. to Sir Jos. Banks P.R.S.', RS Archives, Letters and Papers, Decade XI, No. 55 (dated 11 August 1796).

[51] Bowdler, 'An Account of an Iron Bridge', pp. 6–7.

[52] *Annual Register*, XLIV (1802), 776. They were also granted a patent in 1796. A description of the technique is given in Bowdler's account, pp. 14–16, and in the *Annales des Arts et Manufactures*, II, 166–73.

[53] Bowdler, 'An Account of an Iron Bridge', pp. 24 and 26. As in the case of the other governors, Burdon was not in any great need of the added income, being worth more than £150,000.

[54] *Annual Register*, LXVIII (1827), 270.

[55] *Victoria County History*, Rutland, I, 244.

[56] Earl of Winchilsea, 'Turnips and Grass Compared, as Winter Food', *AY*, XXVI (1796),

with the Earl's husbandry that he gave a detailed description of the Rutland farm, pointing out that Winchilsea's experiments had led to positive results in terms of agricultural production.[57]

We have, then, in the Royal Institution, a completely unexpected situation. Whereas most landowners were not affected by the late eighteenth-century interest in agricultural and industrial expansion, the RI had a large number of Proprietors who were; and its governorship possessed eight of the most active and dedicated improving landlords of the day. In their eyes, 'science' was very much a part of what they were doing: experimenting with new crops or agricultural machinery, opening mines, constructing canals and turnpikes. It was a conception of science which, we have argued, was closely related to perceived economic interests, and it is this conception, rather than Davy's (or later Faraday's), that is the real key to understanding the RI at this time. Davy's work affords the best example of how research at the RI was moulded to fit the ideology of science held by the governors.

The nature of the Proprietors and governors, and the close relations between agriculture and philanthropy, resulted in a public image for the RI during the first decade that was similar to that of the Board of Agriculture, even during 1799–1801, and the governors actively sought to encourage this. Thus the only formal 'advertisement' they placed was in Young's *Annals of Agriculture*, the clearing-house for all agricultural information and the one publication that all agricultural enthusiasts read. A glance at the governors listed in the advertisement would leave no doubt in the reader's mind as to the future orientation of the Institution.[58] The RI also received some publicity in *The Commercial and Agricultural Magazine* (as of 1802 simply *The Agricultural Magazine*), which was popular among the landed interest.[59]

An excellent example of the public image of the RI is furnished by the will of one Edward Goat, a provincial landowner and Colonel of the East Suffolk Militia, who died in 1803.[60] Goat apparently believed that the RI was following the usual programme of most agricultural societies in offering premiums, and thus his will reads:

> To New Society of Husbandry &c lately established in Albemarle Street I leave twenty Guineas to be given as a premium to any person who discovers a principle which may lighten the Draught of Oxen to carriages & a second twenty Guineas to

462–3; 'On the Early Sowing of Oats', *AY*, XXIX (1797), 376–8. Young noted that the latter experiment 'would pay the rent of the land at a high rate . . ..'

[57] 'Some notes concerning the Earl of Winchilsea's husbandry, at Burley-on-the-Hill, Rutlandshire', *AY*, XXXII (1799), 351–81. A description of some of Winchilsea's experiments is given in Gazley, *Life of Arthur Young*, p. 381.

[58] *AY*, XXXIII (1799), 239–45. The advertisement is dated 18 May 1799.

[59] *See*, for example, Vol. II (1800), 218, 293–4, and 369. *The British Magazine* also covered the activities of the RI: *see* Vol. I (January to June, 1800), 176–9, 237–8, 273–4, 361–2, 555–8, and Vol. II (July to December, 1800), 139–40 and 230–4.

[60] Also spelled Goate. *See MM*, IV, 44 (18 March 1805), 70 (6 May 1805), and 75–6 (13 May 1805); and RI Archives, Box 15 (old numbering system), Folder 128. *See also* RASE Archives, *Rough Minute Book*, entry for 3 May 1805. The original will (dated 4 August 1802) is located in the Principal Probate Registry, now part of the collection of the Public Record Office.

the Said Society to be given to any one who produces a Design of a machine which may be purchased at a low price for raising of water to the summit of Buildings.[61]

The RI received notice of this in 1805, but since it did not award premiums, it transmitted the notice to the Board of Agriculture. Lord Sheffield, then President of the Board, thanked the RI and informed it that it would carry out the wishes of the deceased as well as 'the wishes of the Royal Institution, with which respectable body, we shall always be happy to carry on the most friendly intercourse'.[62]

'Society of Husbandry' is only part of the early story of the RI, but it is the most important part. As late as 1820, the Institution was still able to view itself in these terms, even though it was by then no longer agriculturally oriented. A letter from Sir John Cox Hippisley to the President of the Board of Agriculture (Lord Hardwicke), dated 9 April 1820, points up this lingering tendency, and sheds light on the activities of the RI during its early years. Noting that Hardwicke 'stands at the head of the Board of Managers [of the RI] at the present year', and that he himself is an Honorary Member of the Board of Agriculture, Hippisley goes on to propose what in effect amounts to a merger of the two institutions, on the grounds that their purposes have for the most part been the same.[63] From the letter it is clear that the RI is once again in financial trouble, and this may have been Hippisley's motive for writing. After reminding Hardwicke that Professor John Millington delivered a series of lectures in 1819 'On the Application of Mechanics to the Purposes of Agriculture & Husbandry', and that the RI was now considering the establishment of a permanent professorship of botany, he goes on to point out

> how many of the recited objects [of the RI] are congenial with the natural investigations of the Board of Agriculture:—such as the analysis of Soils—of Manures—of Minerals, and other productions of the Earth—the examination and application of Machinery to the purposes of Agriculture . . .,

and he concludes with a proposal 'of practically consolidating, in some respect, the objects of either Institution with manifest advantage to both'.[64] This would

---

[61] The idea of using oxen for draught was a popular one. In 1796 William Wilberforce sent out a questionnaire to various people which formed the basis of his later work on the relief of the poor. Question 22 reads: 'Would it not be politic to encourage the use of oxen in agriculture, draught, &c. in place of that of horses?' (Robert and Samuel Wilberforce, *Life of William Wilberforce* [5 vols.; London: John Murray, 1838], II, 416) William Cobbett also recommended this in his *Treatise on Cobbett's Corn*. Hydraulics was used in agriculture and of course in mining. Goat apparently has second thoughts, as his offer of the second twenty guineas was crossed out, with a note in the margin: 'too much patriotism, my heirs might think'.
An extract of the will was delivered to the RI Printer, William Savage, by the executor of the will, Mr Duverne. It is identical except for minor alterations.

[62] RI Archives, Box 15 (old numbering system), Folder 128.

[63] RI Archives, Box File XIV, Folder 132. The letter is a copy of the original, which has been lost. I have dropped Hippisley's italics in all of the excerpts quoted below, as they are arbitrary and include more than half of the words. Lord Hardwicke became a Proprietor of the RI early in 1800; *see* the letter from Rumford to Hardwicke in BM Add. MS. 35643, ff. 350–1 (dated 1 December 1799).

[64] ibid., pp. 4–5. John Millington (1779–1868), a civil engineer, was Professor of Mechanics at

entail the Board moving from its present location (Sackville Street, a few blocks away) to the house next to the RI; holding the lectures of the Board in the Theatre of the RI; and constructing a passage from the Board's quarters to the Theatre as well as to the Laboratory,

> for the more ready purposes of the analysis of Soils Manures, Minerals &c. and also [providing access] to the Geological Collection and Repository of the Institution, where there are many useful Models of Agricultural Machinery which have occasionally been the subject of Lectures, not less appropriate to those of the Board of Agriculture, than of the Royal Institution.[65]

Naturally, he continues, the Board would permit members of the RI to attend its lectures. The result of all this, he believes, is that the government might be induced to provide

> a small Annual grant—say not exceeding £1,000 ... which would amply insure the permanent prosperity of this eminently useful and national Institution.[66]

The merger did not, of course, take place, and the Board of Agriculture ceased to exist within two years of the letter; but this document testifies to the similarity of interests of the two organizations during the early period.[67] This similarity goes far beyond the items cited by Hippisley. In January of 1800 the 'Committee of Papers &c.' of the Board, which was composed of six members half of whom were RI Proprietors (Langford Millington, John Sinclair and the Earl of Winchilsea), recommended to the Board that it formally adopt a series of investigations, most of which it had already been pursuing in a desultory fashion. These included the condition of the poor, specifically the building of cottages, prevention of famine and finding of a leather substitute;[68] the culture of plants and improvement of wastelands; and the conducting of chemical and practical experiments.[69] Of course, the Board was no more successful than the RI in being able to deal with rural poverty, but in some areas its special relationship to the Institution did help it achieve its goals. Chambers and Mingay regard the Board as a failure, for (among other reasons) 'it had no facilities and but little money for encouraging technical experiments ...'.[70] This judgment is less true, however, if one recognizes that the RI was the Board's laboratory and Davy its salaried employee. The history of the Royal

---

the RI from 1817 to 1829.

[65] ibid., p. 5.

[66] ibid., p. 6.

[67] A note by the clerk who copied the letter out, on p. 7, reveals that Lord Hardwicke approved the proposal, but was replaced as President before his Board acted, and the members of the Board ultimately decided against the merger. *See also* Mitchison, 'The Old Board of Agriculture', p. 65.

[68] A substitute for leather was on the premium list of many agricultural societies, as a means of poviding shoes for the poor at minimal cost. *See*, for example, the Board premium list for 1801 or 1802.

[69] RASE Archives, *Rough Minute Book*, 1799–1801, pp. 15–16 (31 January 1800).

[70] Chambers and Mingay, *Agricultural Revolution*, p. 121.

Institution during the decade of 1802–12 is traditionally identified with Davy, and this is accurate enough; but not for the reasons usually given. As George Eliot suggested in *Middlemarch* (*see* below, p. 61), Davy's national reputation was primarily derived from his authorship of the *Agricultural Chemistry* rather than his decomposition experiments. Davy could well be the perfect example of how, at a scientific institution, structure comes to influence function.

So much has been written on Davy that it might seem gratuitous to attempt yet another portrait of Regency England's most famous scientist. 'No scientist since Newton,' writes J. H. Plumb, 'had so captured the nation's imagination ....'[71] Yet as the foregoing analysis of the RI might suggest, some reassessment of Davy would seem to be in order if we are to understand not only his place in the history of science, but in English social history as well. *Why* did Davy capture the nation's imagination, and what relevance (if any) did this triumph have to the agricultural orientation of the RI? Although the following discussion will, at least partly, be concerned with Davy's personal history, my aim here is not biography per se. For Davy's life, colourful as it is in its own right, more importantly illustrates the process by which certain ideological changes in the uses of science found a public spokesman and a public servant. If Davy electrified his audiences, as his numerous biographers have informed us, it was also the case that he was preaching to the converted.[72]

The choice of Davy as chemical assistant in 1801 still seems puzzling, or at least somewhat accidental. Unlike Young and Garnett, he did not have an established reputation; and R. J. White's remark, that he was 'swept to the pinnacle of fame on a gas-bag',[73] which is actually a minor part of the story, refers only to the period after 1801. Nevertheless, the subject of anaesthesia did experience something of a vogue around the turn of the century, and Davy was at the right place to study pneumatic medicine, viz. Thomas Beddoes' Pneumatic Institution in Bristol.[74] Davy's discovery of the anaesthetic properties of laughing gas (nitrous oxide) did attract some attention. Yet the orientation of the RI was not medical at this time, and it is probably more significant that Davy's name had appeared in print prior to 1801,[75] and that his one lecture at Bristol was on the connection between chemistry and vegetation, a lecture

---

[71] J. H. Plumb, *England in the Eighteenth Century* (Harmondsworth: Penguin Books, 1950), p. 169.

[72] *JAP*; *JD*; John Davy (ed.), *Fragmentary Remains, Literary and Scientific, of Sir Humphry Davy, Bart.* (London: John Churchill, 1858); T. E. Thorpe, *Humphry Davy, poet and philosopher* (New York: Macmillan, 1896); J. G. Crowther, *British Scientists of the Nineteenth Century* (London: Routledge and Kegan Paul, 1935), pp. 3–66; James Kendall, *Humphry Davy: 'Pilot' of Penzance* (London: Faber and Faber, 1954); L. Pearce Williams, 'Humphry Davy', *Scientific American*, CCII (June, 1960), 106–16; Anne Treneer, *The Mercurial Chemist* (London: Methuen, 1963); and Harold Hartley, *Humphry Davy* (London: Thomas Nelson and Sons, 1966). The historiography of Davy is discussed by J. Z Fullmer in 'Davy's Biographers: Notes on Scientific Biography', *Science*, CLV (20 January 1967), 285–91.

[73] R. J. White, 'The Testament of Sir Humphry Davy', *History Today*, III (February 1953), 102.

[74] F. Cartwright, *The English Pioneers of Anaesthesia: Beddoes, Davy, and Hickman* (Bristol: John Wright and Sons, 1952).

[75] Davy's studies on the nature of light, heat, and 'phosoxygen' were published by Beddoes in his *Contributions to Physical and Medical Knowledge, principally from the West of England* (Bristol: Biggs and Cottle, 1799).

(according to one eyewitness) which 'was listened to with uncommon interest ... '.[76] Davy also knew Davies Gilbert (then Giddy), an influential Cornishman who secured his appointment at Bristol and who would later succeed Davy as President of the Royal Society after the latter's death. Whatever the reason for Davy's appointment at the RI, the immediate contact was a man named Thomas Underwood, who knew both Davy and Rumford, became a Proprietor of the Institution, and recommended the young chemist to the Count. Thus began his famous career.

Despite the Bristol lecture, there is no evidence of any real devotion, on Davy's part, to applied science or agricultural chemistry. His first three months at the RI, in terms of interests, were largely his own, and he applied himself to researches on galvanism.[77] Writing to a friend at the Pneumatic Institution during this period, Davy told him: 'I am about 1,000,000 times as much a being of my own volition as at Bristol,'[78] and he had good reason to feel this way. When Rumford hired (or was requested to hire) Davy, he informed him that the Managers wanted him to feel free to continue 'those private philosphical investigations ... by which you have so honourably distinguished yourself and attracted their attention ... '.[79]

All of this was short-lived. Davy hoped to continue his electrochemical researches, but the Managers 'had more utilitarian aims in view for him'.[80] At a meeting of the Managers held on 29 June 1801, it was

> *Resolved*, That a Course of Lectures on the Chemical Principles of the Art of Tanning be given at the Royal Institution, by Mr. Davy, to commence the second of November next, and that respectable Persons of the Trade, who shall be recommended by Proprietors of the Institution be admitted to those lectures gratis.
> *Resolved* farther, That Mr. Davy have permission to absent himself from the Institution during the months of July, August, and September, for the purpose of making himself more intimately acquainted with the practical part of the business of Tanning, in order to prepare himself for giving the above-mentioned Course of Lectures.[81]

The interest of the landowners in tanning had a number of facets to it. Tanning had some importance for contemporary social conditions: agricultural societies were offering premiums to 'the person who shall invent ... the best and cheapest substitute for leather, in the shoes of the labouring poor', and the

[76] RI Archives, Davy Letters, Box 26, Item 92: John Eden to the Rev. J. H. Townsend of Marazion, Cornwall; dated Clifton, 7 January 1833.

[77] Humphry Davy, 'An Account of some Galvanic Combinations, formed by the arrangement of single metallic Plates and Fluids, analogous to the new Galvanic Apparatus of Mr. Volta', *Phil. Trans.*, XCI (1801), 397–402. This was read to the RS on 18 June 1801. Davy also gave a course of lectures on the subject, as well as on pneumatic chemistry.

[78] Quoted in John Davy, *Fragmentary Remains*, p. 64.

[79] RI Archives, Rumford MSS, Letter dated 16 February 1801, copied by Savage.

[80] Hartley, *Humphry Davy*, p. 40.

[81] *MM*, II, 197–8. Also quoted in C. H. Spiers, 'Sir Humphry Davy and the Leather Industry', *Annals of Science*, XXIV (June 1968), 106. It is not clear whether such a course was given. The Index to the *MM* at the RI (there are two volumes, a rough version and a final one), pp. 319ff., lists Davy as giving lectures on tanning and dyeing.

manufacture of leather served as a source of employment for tenants and cottagers.[82] The governors of the RI were undoubtedly interested in its possible use in an allotment programme. In an article Bernard wrote for the SBCP *Reports* in 1799, he stated that the RI should specifically direct itself to the improvement of the 'MEANS OF INDUSTRY ... AMONG THE POOR'. 'Industry in the cottage' was a favourite phrase of his, and he believed it would 'never be effectually obtained, without such an establishment as the institution'.[83] Tanning was obviously an appropriate type of domestic industry, and a scientific study of the subject was also seen as a possible contribution to the problems of rural poverty and social unrest.

There were other reasons for the popular interest in leather. Dr Spiers notes the importance of leather at this time for clothing, transport (harnesses, saddles, and coach chairs), industry (e.g., covering the rollers of spinning machines) as well as other everyday applications.[84] During the last quarter of the eighteenth century the tanning industry grew rapidly. Between 1796 and 1805 the total weight of hides produced amounted to nearly forty million pounds, and the leather industry absorbed most of this.[85] By the end of the century substantial amounts of hides were being imported, and the leather made from these imported hides alone rose from 1.8 million lb in 1772–4 to 4.3 million in 1796–8, approaching twenty million lb by 1810–15.[86] According to Walther Hoffman, the manufacture of leather was the third or fourth most important industry in Great Britain at the beginning of the nineteenth century.[87] In his *Annals of Commerce* David Macpherson ranked it third (after wool and iron) as of 1785, with an annual value of £10.5 million, and second (after wool) by 1805.[88] In such a situation, the RI interest in tanning might have reflected the awareness of a small number of the aristocracy of the importance of industry; and tanning, being so heavily dependent on agricultural products, was an industry in which they could seriously involve themselves. At the very least, the success of the leather industry would increase their rent rolls, since hides

[82] E.g., 'Premiums offered by the Board of Agriculture, for 1802', *AY*, XXXIX (1803), 86; cf. above, footnote 68. From the sixteenth century onward farm workers occasionally obtained employment in local tanneries on a part-time or seasonal basis. *See* L. A. Clarkson, 'The Organization of the English Leather Industry in the Late Sixteenth and Seventeenth Centuries', *The Economic History Review*, Series 2, XIII (1960), 248, and 'Leather Crafts in Tudor and Stuart England', *The Agricultural History Review*, XIV (1966), 25.

[83] 'Extract from an account of the Institution, for applying science to the common purposes of life, so far as it may be expected to affect the poor,' SBCP *Reports*, II, 143–8, dated 1 June 1799.

[84] Spiers, 'Davy and the Leather Industry', p. 99.

[85] Phyllis Deane and W. A. Cole, *British Economic Growth 1688–1959* (2nd ed.; Cambridge University Press, 1967), pp. 57 and 72–3. By way of comparison, the figure for 1751–60 was almost ten million lb less. The leather industry absorbed virtually all the hides from large cattle and about one half the supply of sheepskins annually.

[86] ibid., pp. 73–4.

[87] Walther G. Hoffman, *British Industry 1700–1950*, trans. W. O. Henderson and W. H. Chaloner (Oxford: Basil Blackwell, 1955), p. 282.

[88] The data for 1785 are cited in Cole and Postgate, *Common People*, p. 66. The woollen industries amounted to £17 million in annual value, with iron and its manufactures at about £12 million. The data for 1805 (*Annals of Commerce*, IV, 15) are cited in L. A. Clarkson, 'Organization of the English Leather Industry', p. 245.

and skins came from their own tenant farmers (or from butchers via farmers) who augmented their income by selling these materials to tanners.[89] It should also be added that the industry was, by 1801, in a precarious state. The entire leather-making process revolved around the ability of tannin, an organic extract from the bark of oak trees, to form a precipitate with the gelatin constituent of hides, which makes them water-resistant and impervious to putrefaction. Unfortunately oak bark, which comprised 60% of the tanner's manufacturing cost by 1801, was scarce, and restrictive and outdated manufacturing laws made innovation difficult.[90] The only alternative seemed to be to find a legal substitute for oak bark, a different source of tannin. As the third edition of *The Encyclopaedia Britannica* put it, a 'substitute for oak bark, the price of which has lately been enormous, is the grand *desideratum* in the manufacture of leather'.[91] 'Tanning materials and leather manufacture,' writes Dr Spiers,

> were by no means remote to other members of the Royal Institution, among them [Sir Joseph] Banks, [Samuel] Purkis and [George] Biggin, besides many of the landed aristocracy and gentry who were doubtless concerned with the utilization of the trees on their estates and of the hides and skins from their livestock.[92]

Arthur Young reported that on some estates in Sussex (where the Earl of Egremont had his Petworth estate) the demand for bark between 1782 and 1794 had increased the value of oak by as much as 100%.[93] In Northampton, where the Earl Spencer's estates were located, the leather industry was vital, for a large part of the population supported itself by making shoes. Sir John Sinclair had established a tannery at Caithness very early in his career, and the Duke of Bedford encouraged one tanner, George Biggin, to do research at Woburn.[94] Joseph Banks, whose Lincolnshire farm was near many tanneries, 'was much interested in the exploitation of plants and plant products',[95] and assisted Davy in his tanning research. Philanthropy and profit, or so it seemed, could conveniently go hand in hand.

That the improving landlords would see Davy as an instrument of more profitable estate activity is not at all obvious, nor do I intend a strict Marxist interpretation here. The improvers were immensely wealthy: the Earl Spencer hardly needed to interest himself in leather manufacture to secure his position. Secondly, important as tanning was to the economy, a more crucial issue for agriculture was excessive moisture, and had the RI been simply a stage for economic forces we would expect the landowners to have directed Davy's attention to the problems of drainage. The whole issue of using science is much more complex than the question of economic profit. The landowners tended to

[89] Clarkson, 'Leather Crafts in Tudor and Stuart England', p. 26.
[90] Walter M. Stern, 'Control *v*. Freedom in Leather Production from the Early Seventeenth to the Early Nineteenth Century', *The Guildhall Miscellany*, II (1968), 438–58.
[91] 'Tanning', *The Encyclopaedia Britannica*, 3rd ed., XVIII, Part 1 (1788–97), 308.
[92] Spiers, 'Davy and the Leather Industry', p. 104.
[93] *AY*, XXII (1794), 384.
[94] Handley, 'Sir John Sinclair', p. 8, and Spiers, 'Davy and the Leather Industry', p. 105.
[95] Spiers, 'Davy and the Leather Industry', p. 104.

think in more general terms, and 'science' was something of a catch-all to them. Philanthropy, profit and estate activity, and service to the nation seemed to run together quite naturally; and the landowners' use of science to assist themselves in these matters undoubtedly flattered their image of themselves as active, entrepreneurial, and culturally avant-garde. The problem of moisture was somehow less scientific than tanning, with its obvious link to chemistry; and drainage of the fens had been in progress since the sixteenth century. The new orientation pressed on Davy was thus more a matter of attitude and psychology than financial reward.

Davy had some familiarity with the tanning industry prior to his arrival at the Royal Institution through his friendship with a tanner in Nether Stowey (Somerset), Thomas Poole, who was also a friend of Samuel Taylor Coleridge.[96] Tanners asked Davy, via Coleridge, for help with the theoretical aspect of tanning. 'I understand from Poole,' wrote Coleridge, 'that nothing is so lit[tle] understood as the chemical Theory of Tan[ning], tho' nothing is of more importance in the circle of manufactures.'[97] In reply Davy looked up the work of the French chemist, Armand Seguin,[98] and reported his findings, never realizing that he would soon be called upon to improve the Frenchman's theories. It was thus natural that when required to do so, he should begin with a trip to Poole's tannery. Davy examined the chemical processes of leather manufacture in great detail, and later wrote to his mother that '[w]e are all fellows of the same craft; they are the great practical tanners, and I am a theoretical one'.[99]

Davy spent the last six months of 1801, and much of 1802, working on the problems of tanning. He became friendly with men like Samuel Purkis, at whose tannery he performed many experiments and with whom he toured Wales during part of 1802.[100] His results were communicated to the Royal Society by its President, and published in the *Philosophical Transactions* for 1803.[101] Dr Spiers calls this paper 'a classic in its field', and Hartley referred to it as the 'only noteworthy investigation made by Davy during those early years in Albemarle Street . . .'.[102] For this work, as well as for certain mineralogical researches required by the Managers, Davy received the Copley Medal of the Royal Society in 1805.

Davy's contribution to the tanning industry was two-fold. He improved the work of Seguin and George Biggin, which had advanced the method of deter-

[96] The friendship between Poole, Coleridge, and Davy is a famous one. *See* Mrs Henry Sandford, *Thomas Poole and His Friends* (2 vols.; London: Macmillan, 1888).

[97] Earl L. Griggs (ed.), *Collected Letters of Samuel Taylor Coleridge* (2 vols.; Oxford: Clarendon Press, 1956), I, 591; dated 7 June 1800. The brackets are Griggs'.

[98] Lelièvre and Pelletier, 'Rapport au Comité de Salut public, sur les nouveaux moyens de tanner les cuirs, proposés par le cit. Armand Seguin', *Annales de Chimie*, XX (1797), 15–18.

[99] *DW*, I, 101.

[100] *JAP*, I, 150–5 and 157.

[101] 'An Account of Some Experiments and Observations on the Constituent Parts of Some Astringent Vegetables; and on their Operation in Tanning', reprinted in *DW*, II, 246–87.

[102] Spiers, 'Davy and the Leather Industry', p. 110, and Hartley, *Humphry Davy*, p. 47.

mining the tannin content of various barks by means of precipitating gelatin.[103] Using isinglass as a source of gelatin, a fact noted by Charles Hatchett in 1800, Davy suggested a method of gravimetric analysis whereby all substances could be tested for their tannin content.[104] In this way he 'provided a rational basis for the comparison of the various tannin-containing materials',[105] and did so at a time when many tanners were trying to find substitutes for oak bark. Nor was Davy unaware of the economic significance of his research. In an article aimed specifically at the Proprietors of the RI, which he published in the *Journal of the Royal Institution* for 1803, he outlined a general easy-to-follow recipe for the testing of tannin content which the non-scientist could readily employ.[106] In his 'Discourse Introductory to a Course of Lectures on Chemistry' of 21 January 1802 he emphasized both the simplicity and importance of leather manufacture:

> Tanning and the preparation of leather are chemical processes, which, though extremely simple, are of great importance to society. The modes of impregnating skin with the tanning principle of the vegetable kingdom ... and the methods of preparing it for this impregnation have been reduced to scientific principles.[107]

Even more important was Davy's discovery of a substance that could rescue the tanning industry from its financial straits, and, while not a substitute for leather itself, promised to make shoes for the poor and other leather goods available cheaply because it was an excellent substitute for oak bark. This was catechu, or *Terra japonica*, the extract of a species of mimosa which grows in Bengal and Bombay. It was first suggested to Davy by Sir Joseph Banks, who supplied him with a quantity for examination in December of 1801.[108] In his paper in the *Philosophical Transactions*, Davy discussed the extract in a straightforward and scientific manner, concluding that it contains the largest proportion of tannin of all other substances, and that while four or five pounds of oak bark are necessary for tanning one pound of leather, a half pound of catechu accomplished the same purpose.[109] In the popular article written for the RI journal, however, he made the obvious point:

> Of all the substances that have been examined as to their tanning properties, catechu or terra japonica, is that which is richest in the tanning principle. This substance is the extract of the wood of a species of the mimosa, which grows abundantly in India; and calculating on its price, and on the quantities in which it may be procured, there is great reason to believe that it may be made a valuable article of commerce.[110]

[103] George Biggin, 'Experiments to determine the Quantity of tanning Principle and Gallic Acid, contained in the Bark of various Trees', *Phil. Trans.*, LXXXIX (1799), 259–64.

[104] Charles Hatchett, 'Chemical Experiments on Zoophytes; with some Observations on the Component Parts of Membrane', *Phil. Trans.*, XC (1800), 327–402; and *DW*, II, 250.

[105] Hartley, *Humphry Davy*, p.47.

[106] Reprinted in *DW*, II, 287–96; cf. esp. pp. 293ff.

[107] ibid., p. 317.

[108] ibid., pp. 265–6.

[109] ibid., p. 286. Davy acknowledged his debt to Purkis here, who also did experiments with catechu. In his notebooks (RI Archives, Davy MSS, Vol. 13, notebook c, p. 109) Davy recorded that catechu is about two-thirds tannin, whereas oak bark is less than one-tenth tannin.

[110] *DW*, II, 295.

Writing to Davies Gilbert in October of 1802, he told his friend and mentor that he was wearing a pair of shoes, one in which the leather was tanned with oak bark, the other in which it was tanned with catechu, and that they were equally good. 'We are in great hopes,' he wrote,

> that the East India Company will consent to the importation of this article. One pound of it goes at least as far as nine pounds of oak-bark; and it could certainly be rendered in England for nearly less than four-pence the pound; oak-bark is nearly one penny per pound.[111]

'We', of course, meant the governors of the Royal Institution, and from Davy's letter it is clear that the landowners were gratified with his outstanding success in so short a time. Science did seem to possess the ability to reward (financially or otherwise) those enlightened enough to patronize it. With Davy's work in tanning, science took an important first step at the RI, in a very visible way, toward fitting into its expected role.

The second step along this path was Davy's work in agricultural chemistry. Bence Jones called it an 'accident ... [that] led the thoughts of Davy to agricultural chemistry' in the spring of 1802.[112] In fact, Davy was required to study soils and manures prior to his employment by the Board of Agriculture.[113] Davy's *Syllabus of a Course of Lectures on Chemistry*, dated 5 January 1802, contains an entire section on agriculture which reveals a good familiarity with the scientific aspects of the subject, and which concludes by saying that chemistry is the key to agricultural improvement, a theme he repeated two weeks later in his 'Discourse Introductory ... '.[114] Davy was required to do soil analyses for RI Proprietors independently of the Board of Agriculture, but it was due to the Board that his attention to the subject was rendered systematic.[115] Davy's lectures were intended to fulfil the goals Lord Carrington had in mind for them: 'to combine the results of Science with the

[111] *JAP*, I, 157–8. Davy recommended importation of catechu to the Managers, and the outcome of this is discussed in Chapter 3. *See* RI Archives, Davy MSS, Vol. 2, MS of Lectures 1805–1808, Folder 2, Lecture 3 (vegetable chemistry), twenty-fifth page (the text is not paginated).

[112] *BJ*, p. 201.

[113] Paris records that Davy did an experiment on the growth of oats in 1801 (*JAP*, I, 402).

[114] The section from the *Syllabus* dealing with agriculture can be found in *DW*, II, 410–16; the reference to agriculture in the 'Discourse Introductory' is on p. 315. In addition, it is possible that Davy was required to investigate soil fertility for Thomas Beddoes at the Pneumatic Institution, since it was popularly (and correctly) believed that there was a connection between soil fertility and the gaseous content of the earth. Arthur Young records in his autobiography (Betham-Edwards, *Autobiography of Arthur Young*, p. 150) that he (Young) began doing pneumatic analyses of soil and manure as early as 1783 to determine this relationship, and Beddoes wrote to the Board requesting information on manures in 1798, suggesting that he too was getting interested in the question (RASE Archives, *Rough Minute Book*, 1797–9, loose sheet entitled 'Business for Tuesday March 13, 1798'). This may also account for Davy's lecture on agricultural chemistry while at Bristol.

[115] Cf. Henry Handley, *A Letter to Earl Spencer ... on the Formation of a National Agricultural Institution* (2nd ed.; London: James Ridgway and Sons, 1838), p. 9: Davy's 'vast talents would probably never have been directed to agricultural chemistry, but for the establishment of the Board of Agriculture'.

practical knowledge of Agriculture', making the landowners realize 'the importance of the study of Agriculture as a Science ...'.[116]

At a meeting of the Board of Agriculture held on 27 May 1802, Arthur Young recorded the following decision:

> Resolved —That it be ref[erre]ᵈ to yᶜ Gen[eral] Com[mittee] to consider any & what means to [?] be taken to procure a series of lectures on the applica[tion] of Chem[istry] to Ag[riculture] in the Board Room or with the approb[ation] of the managers of the Royal Institution in their rooms where they already have a laboratory & yᶜ necessary Apparatus and in case such a measure do[es] seem to the Com[mittee] proper, that they do communicate with the Managers of yᶜ Institution for this purpose.[117]

Sir Joseph Banks, a permanent (Official) Member of the Board, was asked to arrange the matter with the RI.[118] The response of the Managers was most agreeable:

> *Resolved*, That the Board of Agriculture be allowed the use of the Lecture Room for the purpose proposed, whenever it shall not be wanted for the regular Lectures of the Institution, provided Subscribers, or persons coming with the tickets of Proprietors, be allowed admission.
> Sir Joseph Banks is requested to make known this Resolution to the Board of Agriculture, and to inform them, that if the Professors of the Institution can be of any service in assisting or forwarding the wishes of the Board of Agriculture in giving these Lectures the Managers have no objections to their being so employed.[119]

Shortly after, Davy contacted the Board on the subject of the proposed lectures, and the Board, being satisfied with his plans for them, instructed Lord Carrington to meet Davy and arrange all the necessary details, so that the lectures could be delivered the following spring.[120] Davy subsequently requested, and received, permission to spend a few weeks of the summer in the country, in order

> that he may be able to collect some information that may be useful in the lectures to be given on Agriculture in the spring, and which may be in other ways connected with the views of the Institution.[121]

He quickly became involved in his new assignment, visiting farmers and becoming acquainted with agricultural methods, analysing different types of

---

[116] 'The SPEECH of the Right Honourable LORD CARRINGTON, delivered at the *Board* of *Agriculture*, on Tuesday, March 15, 1803,' *Communications to the Board of Agriculture*, IV (1805), 231. Carrington was then President of the Board, an RI Proprietor, and the man who hired Davy.

[117] RASE Archives, *Rough Minute Book*, 1801–3, p. 125.

[118] ibid., p. 129.

[119] *MM*, III, 44 (31 May 1802).

[120] RASE Archives, *Rough Minute Book*, 1801–3, pp. 141–2 (8 June 1802), and pp. 153–4 (8 February 1803); *see also* p. 159 (15 February 1803).

[121] Quoted in *BJ*, p. 329.

soils and the chemical effects of manures.[122] Directly upon his engagement with the Board, Sir Thomas Bernard

> allotted him a considerable piece of ground near his villa at Roehampton, where, under his sole direction, numerous experiments were tried, many of which proved highly successful . . ..[123]

Writing to Davies Gilbert in October, he stated:

> I have to deliver a few lectures on Vegetable Substances, and on the Connexion of Chemistry with Vegetable Physiology, before the Board of Agriculture. I have begun some experiments on the powers of soils to absorb moisture, as connected with their fertility.[124]

He also asked Gilbert to furnish him with several pounds of uncultivated soil from his estates.[125] Prior to beginning his first series of lectures Davy did an examination of plants in vegetation,[126] and in 1803 Joseph Banks introduced him to T. A. Knight, a pioneer in plant physiology, with whom he did many experiments.[127] Also previous to these lectures, which were delivered on Tuesdays and Fridays during three weeks in May, 500 copies of Davy's prospectus were printed up for the use of Board members.[128] In the course of these lectures, Davy was granted an expense allowance of £40 for the purchase of lecture apparatus; and upon their completion he was appointed Professor of Chemical Agriculture with a salary of £100 per annum, in addition to the sixty guineas he had already received for the six lectures he delivered. He would also be permitted to publish his lectures for his own profit.[129] On 31 May 1803 Davy was made an Honorary Member of the Board.[130]

With Davy's new title and financial rewards came new responsibilities. At the same meeting at which his professorship was announced, the General Committee of the Board handed down their version of what they expected Davy to do, beyond reading an annual series of lectures. This was

[122] *JD*, I, 339.

[123] *JAP*, I, 188.

[124] ibid., p. 158; dated 26 October 1802.

[125] ibid., p. 159. Gilbert was President of the Penzance Agricultural Society and reported to Arthur Young on his experiments on manure. *See* 'Experiments on use of Sea Salt, as Manure', *AY*, XXVII (1796), 200–1.

[126] RASE Archives, *Rough Minute Book*, 1801–3, p. 159 (15 February 1803).

[127] Treneer, *Mercurial Chemist*, p. 103. Knight was an Honorary Member of the Board of Agriculture. *See* his 'Account of Herefordshire Breeds of Sheep, Cattle, Horses, and Hogs', *Communications to the Board of Agriculture*, II (1800), 172–91.

[128] RASE Archives, *Rough Minute Book*, 1801–3, pp. 232 and 236–7 (29 April and 3 May 1803).

[129] ibid., p. 246 (13 May 1803) and pp. 263–4 (27 May 1803). The figure of £100 was decided upon by Joseph Banks and Lord Carrington's successor, Lord Sheffield.

[130] ibid., 1803–6, p. 1. Davy maintained an active association with the Board at least until 1818. The *Board Minute Books* (these are the full version of the *Rough Minute Books*), III, 90, entry for 17 March 1818, indicates that Davy attended Board meetings eleven times in the year since March 1817, one of the highest attendance records of those present. Records are unavailable for the later period as the *Board Minute Books* for 1808–16 and 1819–22 have been lost, and the *Rough Minute Books* for early 1806 to 1820 are also no longer extant.

8. Board of Agriculture meeting, from Rudolf Ackermann, *Microcosm of London*

to analise such substances as shall be put into the professors hands by y^c Com[mittee] .... it is also necessary that the B[oard] be aware that the Science of agricultural Chemistry is at present in its infancy & that until it is more matured each analysis will take up a considerable portion of the time Mr. Davy can set aside from the duties of his prior engagements; they may however ... be fairly allowed to hope that it will not be long before Mr. Davy with proper assistants under his superintendance will be able to undertake the business of analysing soils, manures &c for individuals wishing to consult him at a moderate fixed price to be paid for the analysis of each substance that shall be put into his hands.[131]

In this way, the RI soon came to serve as an agricultural laboratory. It was not until late in 1805, however, that the Board decided to furnish Davy with a soil analysis laboratory at the Board house proper, which was a short walk from the RI. Davy informed the Board that he would be ready by October 'to analyse any specimen of soil ... properly described relative to situation (level or on a declivity), depth of rain, & subsoil exposure'.[132] By the following year, the laboratory was put to even greater use. At the conclusion of a lecture, the audience adjourned to the Laboratory, where they watched Davy give an experimental demonstration of what he had just lectured upon in theoretical terms.[133]

Davy's agricultural analyses occupied the bulk of his time, at least until 1807, and the extant records give us a clear idea of the nature of his activities. He was almost completely at the beck and call of the landowning class. In the summer of 1803, fortified with a £50 travel allowance from the Board, he conducted a 'soil analysis tour' of England, concluding it with a three-week stay on the estates of Lord Sheffield, the Board President.[134] The Board wanted his studies of soils in printed form, and Davy wrote Arthur Young:

> I shall be able to send you a paper on the Analysis of Soils, with a model of the chemical chest required for the purpose by Tuesday Mor[nin]^g & If you should think the paper worth publication I shall be very glad to see it in the next Volume of the Communications [to the Board of Agriculture] .. ..[135]

Davy's paper was not only published in the *Communications*, but it was considered important enough to be published in pamphlet form;[136] and Davy's work for the Board and the RI motivated John Hippisley, who was then Vice-President of the Bath and West of England Agricultural Society, to urge that

[131] RASE Archives, *Rough Minute Book*, 1801–3, pp. 264–5.
[132] ibid., 1803–6, p. 224 (21 May 1805), and pp. 226–7 (24 May 1805). The entry here suggests that setting up such a laboratory may have required an Act of Parliament, perhaps since the Board was a semi-official body. According to Gazley (*Life of Arthur Young*, p. 489n), a separate laboratory was finally not set up at the Board, but the Board records seem to suggest otherwise.
[133] RASE Archives, *Board Minute Book*, II, 119 (16 May 1806).
[134] RASE Archives, *Rough Minute Book*, 1803–6, p. 2 (31 May 1803), and p. 9 (3 June 1803); *JAP*, I, 182.
[135] BM Add. MS. 35129. f. 252.
[136] *On the Analysis of Soils, as Connected with their Improvement* (1805). See also, *Communications to the Board of Agriculture*, IV, 302–18. It was also reprinted in Nicholson's *Journal* for 1805.

society to set up a chemical laboratory and a 'Committee of Chemical Research'.[137] The following year it heard lectures by a Dr Archer and a Dr Boyd. Boyd conducted soil analyses for landowners using methods outlined by Davy.[138] Davy's laboratory notebook at the RI indicates that he was still working on the analysis of soils for the President of the RI and others in 1806.[139]

Examples of other agricultural researches done by Davy during this period are numerous. As a frequent visitor to the Duke of Bedford's estate at Woburn, he attempted to assess the feeding value of the ninety-seven types of grasses grown there.[140] In 1804 he analysed wheat specimens for Sir Joseph Banks, and did the same for the Board of Agriculture the following year.[141] In 1806 he did studies of paring and burning, clay and marl, and in 1807 reported to the Board on the manufacture of artificial manure.[142] One of his duties was to rule on entries submitted for the premiums of the Board in categories such as salt, manure, marl, and other subjects.[143] Another project was drawing up 'a Statement of his observations on the causes of the Destruction of Young Turnips by the Fly, and the various measures which are most likely to obviate the same ...', as well as a report on turnip seeds.[144] One of the more amusing incidents that occurred at this time involved a 'grow-fast' mixture which Davy prepared for the use of landowners. Writing to Arthur Young, Lord Egremont told him:

> I was fool enough to send to Mr Accum the Chymist ... & get six Gallons of Mr Davies mixture to accelerate vegetation, & I, steeped all my Turnip seed in it & the consequence is that not one seed has vegetated & I have the trouble of sowing a hundred acres over again.[145]

Understandably anxious, Davy replied directly to Young as well as to Sir John Sinclair, saying that if

the mixture be made in the way I have recommended there can be no disagre[e]able

---

[137] Harrison, *The Board of Agriculture*, p. 129; E. John Russell, *A History of Agricultural Science in Great Britain* (London: George Allen and Unwin, 1966), pp. 59–60.

[138] Russell, *History of Agricultural Science*, p. 60.

[139] RI Archives, Davy MSS, Vol. 6, *Laboratory Notebook*. 1805–9, p. 13 (13 May 1806).

[140] Russell, *History of Agricultural Science*, p. 70; *JAP*, I, 181–2. The list of grasses can be found in the RI Archives, Davy MSS, *Common Place Book*, pp. 246–9. In 1807 he was also working on grasses (*Board Minute Book*, II, 167–8, 10 March 1807).

[141] RI Archives, Davy MSS, Vol. 2, Folder 3 (lectures for 1807), 'Veg. Chem. 1807', eleventh and twelfth pages; RASE Archives, *Rough Minute Book*, 1803–6, p. 154 (5 March 1805).

[142] RI Archives, Davy MSS, Vol. 6, *Laboratory Notebook*, 1805–9, p. 14, and RASE Archives, *Board Minute Book*, II, 153 (17 February 1807).

[143] RASE Archives, *Board Minute Book*, II, 134 (6 June 1806), and *Rough Minute Book*, 1803–6, p. 117 (25 January 1805).

[144] RASE Archives, *Board Minute Book*, II, 118 (13 May 1806) and 154 (20 February 1807). Davy's recommendation was to apply chlorine to the turnips, and his results were printed and circulated at once. He also recommended using urine and lime (pp. 121–2, 20 May 1806). According to Peter Mathias, turnips 'were probably the most important single new feed crop which increased the number of sheep ...' (*First Industrial Nation*, p. 76).

[145] BM Add. MS.35129, f. 350, dated 22 July 1806. The excerpt is reproduced in Gazley, *Life of Arthur Young*, p. 503.

consequences—& even a mistake in the process would not be attended with danger —I have made an alteration in the exp[erimen]ʹ which must prevent any misconception or error . . ..[146]

Barring incorrect preparation of the mixture by Davy's chemical assistant, Frederick Accum, we can conclude that agricultural chemistry was still in a fledgling state.[147]

The following year, the Board asked Davy to do an agricultural survey of Cornwall, but he cried off on the grounds of ignorance, offering instead to do a geological and mineralogical study of the county, which he had for all practical purposes already done for the RI between 1804 and 1806. Upon completion of this in 1808, the Board paid him £100.[148]

All these years of theoretical and experimental farming experience, as well as his annual lectures on agriculture, went into the publication of his lectures in 1813.[149] Public reaction was generally favourable during the early years, but the lectures did not actually contribute much to agriculture that was of any real practical value. Davy was unable to deal effectively with the crucial question of the relation of the soil and its constituents to plant growth. He does not appear to have been a vitalist, but this left him no adequate explanation of how inorganic materials or insoluble humus got into plants, or in general how plants obtained their nourishment. Nor was he able to come to terms with the role of nitrogen in plant growth. Davy did improve upon traditional theories of organic manure, discussing the nature of putrefaction and the nutrients released by the process; but the description was highly impressionistic, and as the value of organic manures had been known since antiquity it could hardly have made any difference for empirical practice.[150] These limitations, even in

---

[146] ibid., ff. 346–7. The letter is dated from Killarney, June 15th (1805). Davy toured Ireland and Cornwall in the summer of 1805, and returned to Ireland in 1806.

[147] Frederick Accum (1768–1838) was Assistant Chemical Operator to Davy from March 1801 to 5 September 1803, after which he became a lecturer at the Surrey Institution until 1809. He spent his life imitating Davy in many ways. A consulting chemist for manufacturers, he took resident pupils into his laboratory, designed and sold agricultural chests for landowners, published on soil and ore analysis, and became a director of the new Gas Light and Coke Company in 1812. His students included the future RI governor, the third Lord Palmerston (later Prime Minister), and Benjamin Silliman, who once commented that Davy had been something of a model for Accum. In 1820 Accum caused an uproar with his muckraking *Treatise on the Adulteration of Food and Culinary Poisons*, which so incensed the guilty parties (mentioned by name) that they persuaded the Managers to bring charges against Accum for mutilating library books, whereupon he fled the country. The charges appear to have been justified, although the scholarly practice of removing leaves from books was common enough at the time.

See C. A. Browne, 'The Life and Chemical Services of Frederick Accum,' *Journal of Chemical Education*, II (1925), 829–51, 1008–34, and 1140–9, and the sketch by R. J. Cole in *Annals of Science*, VII (1951), 128–43.

[148] Letter from Davy to Young, 30 April 1807, BM Add. MS. 35129, ff. 400–1, and RASE Archives, *Board Minute Book*, II, 231 (18 March 1808). *See also* RI Archives, Davy MSS, Vol. 15, notebook h, labelled '*Geology of Cornwall*', and his 'Hints on the Geology of Cornwall' in *DW*, VI, 193–203.

[149] Humphry Davy, *Elements of Agricultural Chemistry, in a Course of Lectures for the Board of Agriculture* (London: W. Bulmer, 1813).

[150] Margaret Rossiter provides a good summary of Davy's work on pp. 11–19 of *The Emergence of Agricultural Science: Justus Liebig and the Americans, 1840–1880* (New Haven:

contemporary terms, throw the popularity of the book into sharp relief. In fact, its objective utility was irrelevant, for it was tremendously important in the eyes of farmers and landowners. It was exactly what the times required, and went through six English and two American editions by 1839 and was translated into French, German, and Italian. As the definitive work on agricultural chemistry for the next thirty or forty years, it was possibly the most widely read guide to agriculture ever written.[151] Even Liebig, who began to discredit much of Davy's work in the 1840s, recognized the significance of that work when he wrote that he had 'endeavoured to follow the path laid out by Sir Humphry Davy ... '.[152]

The legacy of Davy for agriculture was captured by George Eliot in *Middlemarch* (Ch. 2), in which she effectively portrayed him as having become identified, throughout the countryside, with the scientific approach to the subject. Thus Sir James Chettam tells Dorothea that he is

reading the *Agricultural Chemistry* ... because I am going to take one of the farms into my own hands, and see if something canot be done in setting a good pattern of farming among my tenants.

Dorothea's father objects to scientific farming, 'electrifying your land ... and making a parlour of your cowhouse', but Dorothea sees it as a means of making the most of one's land, infinitely preferable, she exclaims, to the keeping of horses and dogs.

With agricultural chemistry, as with tanning, Davy made a profound impact on farming circles, not so much in actual scientific contribution as in reinforcing the belief that science was something in which one could invest. It is too simplistic to state that Davy made scientific farming fashionable. Rather, Davy was the vehicle by which science was channelled into a specifically agricultural direction, a direction in which (with the exception of one lecture at Bristol) he had not previously evinced any interest. Davy's agricultural labours were a vindication and fulfilment of the reigning ideology. The Royal Institution had become a dramatic proving ground not so much of the idea of science for the 'relief of man's estate', but rather for its enrichment.[153]

Yale University Press, 1975), Davy himself acknowledged the problems with some of his hypotheses in the 1827 edition of the *Agricultural Chemistry*.

[151] Russell, *History of Agricultural Science*, p. 75. The *Agricultural Chemistry* had a very great impact on American husbandry (*see* Wyndham D. Miles, ' "Sir Humphrey Davie, the Prince of Agricultural Chemists" ', *Chymia*, VII [1961], 126–34), and Davy was engaged to repeat his lectures in Ireland before the Farming Society of Ireland and the Dublin Society (*JD*, I, 427; Treneer, *Mercurial Chemist*, p. 116).

[152] Quoted in Russell, *History of Agricultural Science*, p. 98.

[153] English science had very little agricultural application before Davy, and certainly none that was carried out in such a public and influential way. The Royal Society's interest in the subject was short-lived (*see* Reginald Lennard, 'English Agriculture under Charles II: The Evidence of the Royal Society's "Enquiries" ', *The Economic History Review*, IV [1932], 23–45). Random studies were done by John Evelyn, Nehemiah Grew, John Mayow, Stephen Hales, *et al.*, but their value to agriculture (even in contemporary terms) is questionable, and they had no impact on farming circles. Davy did, however, benefit from reading Lord Dundonald's *Treatise shewing the intimate connexion that exists between chemistry and agriculture* (1795).

In 1804 a third topic was selected for Davy's expert investigation, and as with the previous ones, this was not accidental. Davy himself advanced the thesis that a 'general knowledge ... of geology [is] essential to the scientific agriculturalist,'[154] but as we have already noted, entrepreneurial activity was not confined to agriculture. For many of the improving landlords, mining was a natural part of estate activity, and it was in this area that geology had its critical application. We shall discuss the formation of the Mineralogical Collection more fully in Chapter 3, since the story pinpoints the discrepancy that had arisen between science for public vs. private use, and deserves a separate treatment of its own; but our interest at this point is in illustrating the impact of this development on Davy. By 14 May 1804 a proposal was formally drawn up to raise £4,000 by subscription,

> for the purpose of forming at the Royal Institution a Scientific Collection of Minerals under the arrangement of an eminent Mineralogist, and of establishing an additional Laboratory there, under the direction of an eminent Chemist; to be exclusively employed in the assay of metals, and in the advancement of mineralogy and Metallurgy.[155]

As had been the case with his agricultural investigations, Davy was quick to seize the opportunity to prove the value of science to its patrons. He had inserted a rather elementary section on metallurgy in his *Syllabus of a Course of Lectures on Chemistry* (1802),[156] but it was not until the above decision that he turned to the subject in earnest, and as his notebooks indicate, in depth. In June of 1804 the Managers granted Davy a leave of absence, for making 'a tour into the north of Britain for the purpose of collecting Minerals, and of gaining information on the Subjects of Geology and Agriculture'.[157] Davy spent the summer collecting minerals in England and Scotland and making notes and sketches of places like a sandstone quarry near Glasgow. On all of his trips, Davy took time to record the geological constituents of the areas he visited; and his diagrams and sketches, carefully done and often beautiful, occasionally contain labels of ores or minerals on various parts of the landscape (*see* Plate 9b). Upon his return he prepared a series of lectures on '*Geology, or the Chemical History of the Earth*'.[158] His lectures provided the Proprietors with detailed information on copper mining, the strata of different areas, and the techniques of uncovering veins of various metals and minerals. In 1806 he added a series of evening lectures on geology to his repertoire.[159] Between 1805 and 1806 the RI paid Davy to make two mineralogical tours of Ireland and a trip to Cornwall, on the latter of which he was accompanied by Thomas Bernard.[160] An

---

[154] *JAP*, I, 385.

[155] *MM*, III, 278 (14 May 1804). The RI agreed to pay up to £1,000 for any additional expense connected with the project.

[156] *DW*, II, 428–31.

[157] *MM*, III, 304 (18 June 1804).

[158] ibid., IV, 8–9 (11 January 1805). 'Sketch of the Sandstone Quarry near Glasgow' is in the RI Archives, Davy MSS, Vol. 15, notebook e, p. 146 and *passim*.

[159] *MM*, IV, 129 (30 December 1805).

[160] *JD*, I, 269–70 and 274ff.; *MM*, IV, 183–4 (19 May 1806). Davy's journal of his second trip

9a. Frederick Accum

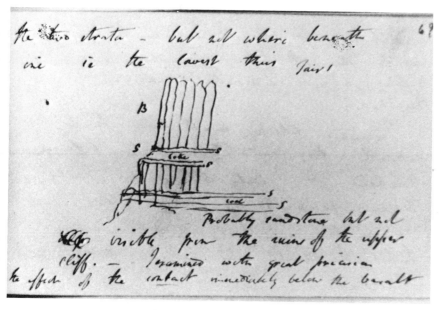

9b. Sketch of a landscape
by Davy

10. 'Sheep Shearing at Woburn,' painted by George Garrard in 1804

examination of his extant notebooks reveals that the time spent on mineralogy was considerable, although there is obviously no way of knowing how many private analyses he undertook at this time.[161] 'All the specimens necessary for elucidating the minute characters of geology,' Davy boasted, 'are to be found in the museum of this Institution ....'[162] Once again it becomes clear how the governors' conception of science changed the direction of scientific research.

It was by such active and successful participation in the plans of the governors that Davy acquired his strong position at the Institution. It was not merely a case of 'selling' science to his audience: the governors, at least, were sold on its potential value before the RI was founded. What Davy was able to do in the Laboratory was justify their investment, verify their belief in science by making their expectations more precise. 'Selling' science to its patrons meant demonstrating in concrete terms exactly what science could do. Here, right here, was catechu, that made excellent shoes at a fraction of the usual price. Here too were the Board lectures, the *Agricultural Chemistry*, the Mineralogical Collection, the lectures on geology, and ultimately the miners' safety lamp. Almost 'all of Davy's discoveries', notes Professor Plumb, 'had the same obvious, practical value for an alert industrial society ....'[163]

Much emphasis has been placed on Davy's personal charm as the key to his success, and he was certainly gifted in that way. Yet a deeper understanding of his rise to national prominence may be acquired by examining the content of his lectures, rather than just their style. They were, to say the least, permeated by the ideology of science that had made Davy's success possible in the first place. The significance of geology to mining, for example, was a theme Davy reiterated constantly. He used examples such as copper mining in Cornwall and Somerset to show how great losses had been incurred by miners ignorant of this knowledge.[164] In one lecture in 1805, he told his audience that coal and metallic veins were often found in secondary strata, and then described these in detail. In the same lecture, he talked about coal and basalt, giving his listeners easy methods of recognizing the latter and testing the quality of the former.[165] Frequently he discussed the strata of specific areas owned by the Proprietors

---

to Ireland is published in *DW*, VII, 146–68.

[161] E.g., a letter to Gilbert, 18 May 1806, preserved at the Royal Institution of Cornwall (Truro), describes the results of 'a laborious series of experiments' on the quantity of iron in copper specimens, done for a group of men who had approached Gilbert with the problem. I am grateful to Mr H. L. Douch, the Curator, for providing me with a transcript of this letter. *See* RI Archives, Davy MSS, Vol. 15, notebooks a,b,c,d,e,g,h,i, and Vols. 16 and 17, for Davy's geological research.

[162] *DW*, VIII, 204; from a lecture of 1805.

[163] Plumb, *England in the Eighteenth Century*, p. 169.

[164] 'Introductory Geological Lecture,' in *DW*, VIII, 196–7. This lecture was given in 1811, but John Davy tells us it was based on a lecture of 1805. Davy was familiar with Somerset mines because Thomas Poole owned one and consulted Davy about it. A letter to Poole on the subject reveals that by 1805 Davy had acquired a great deal of knowledge about metallic deposits on the basis of strata and shifts. It also reveals he was familiar with mining in Alston Moor, Derbyshire, and Staffordshire (*JAP*, I, 211–12).

[165] RI Archives, Davy MSS, Vol. 17, *Lectures on Geology 1805*, 'Geology Lect. 6', twentieth page ff. and thirty-eighth page to fifty-second page.

themselves.[166] In another lecture of 1805, Davy elaborated on metallurgy, giving a straightforward account of how to use geology to detect metallic veins: '*Primitive limestone* has sometimes been discovered productive of copper—And in *the Quartz & calcareous veins that it contains*, Gold and silver occur—'.[167] '*Sandstone* seldom or never contains *any metals*—Metallic Veins are never found in loose shale nor in *Columnar Basalt*—.' He went on to provide his audience with 'certain general Rules with regard to the indications of metals which will apply equally to the Veins in the primitive & secondary Rocks—'. Occasionally, Davy used drawings or coloured maps, for example to point out 'the general direction of the principal Veins in Cornwall containing ore'. And, as he often did in his lectures on agriculture, he appealed to simplicity: 'No branch of *practical Knowledge* can be more simple or more intelligible than the History of the position & arrangement of metallic Veins ...'. By 1810 the Dublin society was paying him £750 to lecture on geology.[168]

It was, however, in the field of chemistry that Davy best fulfilled his role as (to use his own term) a 'campaigner' in the cause of science.[169] In his introductory chemistry lecture of 1802, for example, he told his audience:

> The knowledge of the composition of soils, of the food of vegetables, of the modes in which their products must be treated, so as to become fit for the nourishment of animals, is essential to the cultivation of land; and [the landowner's] exertions are profitable and useful to society, in proportion as he is more of a chemical philosopher.[170]

'Science has done much for man,' he continued, 'but it is capable of doing still more; its sources of improvement are not yet exhausted ....'[171] 'Whosoever reasons upon agriculture,' he told the Board on one occasion,

> is obliged to recur to this science. He feels that it is scarcely possible to advance a step without it ..... If a person journeying in the night wishes to avoid being led astray by the ignis fatuus, the most secure method is to carry a lamp in his own hand.[172]

Davy made the prospect of becoming a part-time scientist an exciting one. He conjured up before his audience the image of the enlightened landowner, riding through his fields: 'a small portable apparatus; a few phials, a few acids, a lamp and a crucible are all that are necessary'.[173] Davy was no Euclid, telling the King there was no royal road to geometry. Rather, he allayed the landowners' fears by emphasizing the simplicity of science:

---

[166] E.g., ibid., fifty-second page, discusses the coal and other strata near Durham, where the Bishop of Durham, Rowland Burdon, and other Proprietors owned land.

[167] ibid., thirty-first page; italics Davy's. All of the following quotations are from this lecture; *see* the thirty-third, thirty-fourth ff., thirty-ninth, and forty-fifth pages.

[168] *JD*, I, 428–9.

[169] RI Archives, Davy MSS, Vol. 22, notebook a, p. 41.

[170] *DW*, II, 316.

[171] ibid., p. 319.

[172] *Agricultural Chemistry*, p. 23.

[173] ibid., p. 25.

[I]t is fortunate for the extension of useful discovery that this *connection* [between agricultural and chemistry] is founded upon very simple *principles* which may be *easily acquired* & which are almost independent of the more abstruse and more complicated parts of the Science.[174]

But the key weapon in his arsenal was the technique of presenting science as the mainspring of social progress, and praising those who patronized it:

The arts and sciences [he told his audience] . . . are in a high degree cultivated, and patronized by the rich and privileged orders. The guardians of civilization and of refinement, the most powerful and respected part of society, are daily growing more attentive to the realities of life; and, giving up many of their unnecessary enjoyments in consequence of the desire to be useful, are becoming the friends and protectors of the labouring part of the community. The unequal division of property and of labour, the difference of rank and condition amongst mankind, are the sources of power in civilized life, its moving causes, and even its very soul; and in considering and hoping that the human species is capable of becoming more enlightened and more happy, we can only expect that the great whole of society should be ultimately connected together by means of knowledge and the useful arts; that they should act as the children and one great parent, with one determinate end, so that no power may be rendered useless, no exertions thrown away. In this view we do not look to distant ages . . . [but instead] look for a time that we may reasonably expect, for a bright day of which we already behold the dawn.[175]

The effect of such words upon an avant-garde upper class can easily be imagined. It produced, John Paris tells us, an 'extraordinary sensation', and copies of the talk had to be printed up for the personal possession of the Proprietors.[176]

A rough draft of another speech delivered at the RI also provides insight into this method of appealing to the audience.

It is fortunate [wrote Davy] that those classes of society which contain the natural Patrons of improvement likewise contain persons who have the most leisure as well as the amplest means of promoting its cultivation. The loftier the source of a stream the more rapid, copious, and pure are its waters.[177] There have been and there are noble examples of men of Rank[178] . . . who have assisted by their influence and fortunes & their example the advancement of Science & the useful arts, and I trust the number will increase. *Such* Benefactors of society have been always the greatest favourites with the public . . . . It is from the higher classes that the tone and spirit is derived which passes into the middling and lower classes[179] orders of society.[180]

[174] This is from the fifth page of the MS copy (RI Archives, Davy MSS, Vol. 21, notebook b) of Davy's lectures on agriculture, copied out by a clerk. The italics are presumably for emphasis in oral delivery.

[175] *DW*, II, 323.

[176] *JAP*, I, 131–3.

[177] This last sentence was crossed out.

[178] Written between the lines at this point, and then crossed out, is the following: 'I am proud to refer to the example [first version] has been set in this room [second version] displayed in this room'.

[179] 'Classes' crossed out.

[180] RI Archives, Davy MSS, Vol. 1, Miscellaneous Lectures, forty-seventh and forty-eighth pages.

This was a theme repeated in his lectures to the Board of Agriculture:

> It is from the higher classes of the community, from the proprietors of land; those who are fitted by their education to form enlightened plans, and by their fortunes to carry such plans into execution; it is from these that the principles of improvement must flow to the labouring classes of the community; and in all cases the benefit is must be [*sic*] always likewise the interest of the proprietors of the soil.[181]

Davy's praise of the 'enlightened' enabled him to secure some of the patronage he sought for science, whether in the form of subscriptions to his published works, his salary, or funds for travel, apparatus, and research. 'Shall Englishmen slumber in that path which [Newton and Bacon] have opened,' he asked his audience,

> and be overtaken by their neighbours? Say, rather, that all assistance shall be given to their efforts; that they shall be attended to, encouraged, and supported.[182]

National glory was a useful theme, but economic interest was not without its attractions. In 1809, when Davy wanted the governors to build a larger voltaic pile for his electrochemical researches, he knew exactly how to couch his appeal for funds: electricity, he said, is a contributing factor to soil fertility.[183] The cost of the new battery was heavily subscribed to by the governors and Proprietors.[184]

It is difficult to recapture the excitement that Davy generated over science at this time. By 1804 the demand on the part of the Proprietors to watch Davy actually at work was so great that one of the Laboratory walls had to be torn down to accommodate spectators.[185] Davy became a regular visitor at the agricultural exhibitions popular at this time, where the great landowners toasted his health.[186] As June Fullmer has written, 'his friends were drawn increasingly from those circles which represented the ruling classes ... '.[187] A famous painting by George Garrard shows Davy at a sheep-shearing at Woburn in 1804, in which T. W. Coke, the famous agriculturalist, is represented conversing with Banks, Sinclair, and Arthur Young, while 'Professor Davy is standing in a listening attitude behind him' (*see* Plate 10).[188]

[181] *Agricultural Chemistry*, p. 24. This paragraph was reprinted in the review of Davy's book in the *Gentleman's Magazine*, LXXXIV (1814), Part 1, 466–8. The supposedly happy interaction of social classes was a common theme at the time; *see*, e.g., Thomas Bernard in the SBCP *Reports*, II, 25–6.

[182] *DW*, VIII, 359.

[183] RI Archives, Davy MSS, Vol. 3, Folder 1, labelled 'Lecture 4, El. Chem *1809*', thirty-fifth page. Davy probably believed this: see Russell Garnier, *History of the English Landed Interest* (2nd ed.; 2 vols.; London: Swan and Sonnenschein, 1908), II, 430.

[184] RI Archives, Box File IIA, Item 81, *Subscription for constructing A Voltaic Apparatus on a Great Scale for pursuing New Researches in Chemistry & Natural Philosophy.*

[185] *VM*, 19 April 1804; *MM*, 4 June 1804.

[186] *JAP*, I, 181n–182n.

[187] J. Z. Fullmer, 'Humphry Davy's Adversaries', *Chymia*, VIII (1962), 147. Davy was, for example, regularly a part of the Spencers' Christmas circle at Althorp. See Thomas F. Dibdin, *Reminiscences of a Literary Life* (2 vols.; London: John Major, 1836), I, 228n.

[188] A. M. W. Stirling, *Coke of Norfolk and His Friends* (2 vols.; London : John Lane The

11. Davy as a young man, 1804

At the sheep shearing at Holkham in 1808, T. W. Coke referred to 'a friend of mine, who sits near me, Professor Davy ... upon whose judgment on account of his extensive chemical as well as other scientific knowledge, I place the highest reliance ...'.[189] There is some justice, then, in Sir John Sinclair's claim that Davy's glory was due to his association with the Board of Agriculture:

> No circumstance could be more gratifying to me, than to have a share in promoting the success of so eminent a philosopher as Sir Humphry Davy, whose talents have thrown such a lustre on his age and country. To his more early friends, Sir Joseph Banks and Mr. Davies Gilbert, he was greatly indebted; but *no circumstance contributed more to his success in life, than his connection with the Board of Agriculture*, as he derived considerable emolument from his services to the Board,* and there became acquainted with a number of the most distinguished characters in the kingdom.[190]

This was how Davy became the 'lion' of English society. His career at the RI was not just the voltaic pile and isolation of the alkaline metals; not just the popular lecturer exciting the social elite of London; but he was more importantly the author of the *Agricultural Chemistry*, the superintendent of the Mineralogical Collection, and the investigator of tanning, and this Davy was much more true to and representative of the aims and objectives of the early RI. He absorbed the ideology of science of the improving landlords and reinforced their belief in it at every opportunity. He became, without parallel in modern history, England's high priest of applied science.

The full impact of the agrarian changes of the late eighteenth century upon English science now becomes clear. With a concentration of atypical improvers at the RI, the stage was set for the institutional emergence of an ideological alternative to the dilettanti conception of science. For the flurry of scientific activity in the North—the clubs, coffee houses, and diffusion of technology referred to earlier in this chapter—remained, despite its obvious energy, inchoate. The formation of scientific discussion groups was sporadic and fitful, the groups normally unrelated and frequently short-lived. They were a reaction to circumstances rather than a deliberate attempt to shape them. They generated what Charles Gillispie has termed a 'natural history of industry', and the affinity to the amateur tradition is obvious.[191] Gears and engines take the place of voyages and shells in the catalogue of description, but the actual bending of this knowledge to economic ends remains essentially unattempted, despite the popularity of the idea. Much of the motivation for this activity, as I have argued elsewhere,[192] was the desire to be aristocratic; to gain access to the Royal Society, rather than rival it. This whole movement, of which the Lunar

Bodley Head, 1907), I, 438n.

[189] *JAP*, I, 182n.

[190] John Sinclair (ed.), *The Correspondence of the Right Honourable Sir John Sinclair, Bart.* (2 vols.; London: Henry Colburn and Richard Bentley, 1831), I, 431; italics mine. The footnote to the asterisk reads: 'Besides £100 *per annum* for lecturing, he received for his lectures £1,000 from a bookseller'.

[191] C. C. Gillispie, 'The Natural History of Industry', *Isis*, XLVIII (1957), 398–407.

[192] *See* the Introduction, footnote 14.

Society of Birmingham is probably the most famous representative, possessed no central focus, and did not—indeed could not—amount to the beginnings of modern scientific organization.[193] What the Baconian vision really required to become a truly legitimate rival to the older ideology was support on the part of the very class with which the dilettanti tradition was invariably associated. There had to be, in short, a departure within the ruling class itself from traditional notions, and it is this development that makes the Royal Institution a turning point in the organization of science. Although it never possessed a research *staff* as such, the RI came much closer to the picture of Salomon's House envisaged by Bacon than any other English scientific organization to date. It was thus in the first decade of the nineteenth century that the gentleman amateur tradition was challenged in an institutional way. The notion of arranging laboratory research, even partly, for the express purpose of economic enterprise was (at least theoretically) a complete antithesis to the hegemonic tradition. Despite his own interests, Davy became the embodiment of this entrepreneurial ideology, validating it in the Laboratory and propagandizing for it in the Lecture Theatre. Thus science entered the nation's life in the form of a commodity, available to those sources of wealth which found it convenient to their purposes. Numerous institutions deliberately copied the RI's format, and certainly its ideology, as a model for scientific organization.[194]

We have, finally, to conclude our examination of Davy's career by briefly touching upon the influence of his institutional role on his actual scientific talents. If Davy became the high priest of applied science, it was also the case, as Coleridge astutely remarked in 1801, that 'chemistry tends in its present state to turn its Priests into Sacrifices'.[195] Davy obtained for science some of the patronage he sought, but paid the price of his own creativity. 'In the face of ... evidence that Davy was intellectually capable of significant theoretical work,' writes Robert Siegfried, 'we are obliged to ask why he did not achieve success in this area of science.'[196] Davy's earliest work, while still with Beddoes

[193] R. E. Schofield claimed that the Lunar Society was a 'pilot project or advance guard of the Industrial Revolution', but it remains a dubious argument (*Lunar Society of Birmingham* [Oxford: Clarendon Press, 1963], p. 439). *See*, however, Neil McKendrick, 'The Role of Science in the Industrial Revolution: a Study of Josiah Wedgwood as Scientist and Industrial Chemist', in M. Teich and R. Young (eds.), *Changing Perspectives in the History of Science* (London: Heinemann Educational, 1973), pp. 274–319.

[194] Although the Royal Institution is the most dramatic illustration of the role of agricultural innovation in the creation of organized support for scientific research, it is probably not unique. In his study of late eighteenth and early nineteenth-century agricultural societies, Kenneth Hudson almost inadvertently demonstrates that sponsorship of scientific activity frequently had an agrarian impulse behind it, and that 'science' and 'agriculture' were often synonymous (*Patriotism with Profit: British Agricultural Societies in the Eighteenth and Nineteenth Centuries* [London: Hugh Evelyn, 1972]). The Dublin Society, for example, had a laboratory and hosted lectures on experimental philosophy. Davy's cousin Edmund served as professor there for a time and repeated the pattern of Sir Humphry's career—discovering acetylene in-between his investigations of potato starch. Significant as the RI was in getting science publicly organized on an entrepreneurial basis, Hudson's survey may suggest that it was part of a larger pattern, though certainly the most outstanding example of this phenomenon.

[195] Griggs, *Coleridge Letters*, II, 768; Coleridge to Robert Southey, 21 October 1801.

[196] Robert Siegfried, 'The Chemical Philosophy of Humphry Davy', *Chymia*, V (1959), 193–201.

at Bristol, was theoretical, and his first three months at the RI devoted to investigations in electrochemistry. In these, Davy was pursuing a suggestion of Davies Gilbert that an analysis of galvanism would ultimately reveal the nature of heat and light. He wrote to Gilbert during this time that he was conducting research with a view to 'destroying the mysterious veil which Nature has thrown over the operations and properties of etherial fluids . . .'.[197] Davy's Bakerian lectures, and his *Elements of Chemical Philosophy* (1812), contain much in the way of tentative speculation, and suggest a constant puzzling over theoretical questions.[198] Yet Davy never succeeded in establishing a comprehensive theoretical structure for chemical science, largely because he literally had no time to think during his formative years. Thus Hartley writes:

> In his early years at the Royal Institution Davy was not altogether free to develop his own ideas, as a committee decided what researches should be undertaken by the Professor. First, there was tanning and then in 1804 the Managers decided to form a collection of minerals and to set up an assay office for the improvement of mineralogy and metallurgy. So Davy had to turn his head to mineral analysis, which was not his natural bent . . . . It was only in the intervals of his other duties that Davy was able to continue his researches in voltaic electricity until the autumn of 1806 . . ..[199]

Because his electrochemical researches were not of central importance to the governors of the RI, they were not allowed, until later, to assume central importance for him, and even then he was still burdened with many mundane obligations. 'My *real*, my *waking* existence,' he wrote almost plaintively to Thomas Poole, 'is amongst the objects of scientific research,'[200] but until late in 1806 there was little time to devote to theoretical reflection. Hartley rightly concludes that 'it is questionable whether the best interests of [pure] science were served by these distractions'.[201]

The most meaningful way to view Davy, I believe, may be the way in which he viewed himself. Davy's pre- RI days were highly speculative, and his early notebooks reveal that he was asking basic questions about the nature of physical reality. Upon his arrival at the RI, Davy was made to see that if science were to survive in the context of the times, it had to do so on the terms of those wealthy enough to patronize it. In an England of agricultural improvement and nascent industrialization, this meant appealing to science's economic potential. Thus Davy conducted, from 1801 to 1805, 'four campaigns of hard fighting in the Cause of chemistry . . .'.[202] At the first opportunity, however, he returned to his theoretical studies, such as the nature of chemical affinity, and 'to the rather wild speculative attitude of his phosoxygen days', speculations

---

[197] *JAP*, I, 109–10; *see also* pp. 129–30.

[198] Cf. Hartley, *Humphry Davy*, pp. 50–80, and Colin A. Russell, 'The Electrochemical Theory of Sir Humphry Davy', *Annals of Science*, XV (1959), 1–25.

[199] Harold Hartley, 'Sir Humphry. Davy, Bt., P.R.S.', *Proceedings of the Royal Society of London*, Series A, CCLV (1960), 161–2.

[200] *JD*, I, 140. The letter is dated 1 May 1803 and the italics, significantly, are Davy's.

[201] Hartley, *Humphry Davy*, p. 45.

[202] RI Archives, Davy MSS, Vol. 22, notebook a, p. 41.

12. Davy in later years

that occupied much of his time between 1807 and 1810.[203] Despite his public campaigning, Davy did not privately feel that utility per se was the purpose of science. He was far more interested in civilization advancing the cause of science than vice versa. He privately detested men like Accum and Brande, who used their knowledge to obtain personal profit,[204] and may have seen utility in general as a rather shallow goal, as the following private memorandum suggests:

> The scientific taste in America seems to me almost always connected with gross profit or common utility ... in the higher departments of the profound & refined branches of science they have little or nothing ....[205]

In this sense, the traditional interpretation of Davy is correct: he *was* a romantic, a natural philosopher, a theoretician by desire and temperament. In terms of his actual career, however, he was a technological scientist, and his proper place in the history of science is not that of an intellectual, but of an intellectual entrepreneur. His career marks the first serious attempt to turn the fragmentary and virtuoso character of English science into an enterprise wealthy enough to support philosophers and romantics. Davy was, as Coleridge wrote to Poole, 'determined to mould himself upon the Age in order to make the Age mould itself upon him'.[206] It was the practical tasks at the RI, and the organization of that institution, that suggested to him the possibilities of such a moulding in the first place, but they also frustrated him from becoming the type of thinker he ardently wished to be. His speculations never matured beyond a tentative, albeit highly suggestive stage. Davy's impact on the social history of science was enormous, but his significance for the discipline of chemistry itself was best summed up by the Swedish chemist, J. J. Berzelius, in a letter to Wöhler:

> Had Davy ... been forced to occupy himself from the beginning with so much writing as it is now your lot, I am convinced he would have advanced the science of chemistry a full century; but he remains, only a 'brilliant fragment'.[207]

---

[203] Hartley, *Humphry Davy*, p. 60.

[204] *See* Davy's comments in the MS published in J. Z. Fullmer, 'Davy's Sketches of His Contemporaries', *Chymia*, XII (1967), 127–50. He called Frederick Accum, whose gas lighting scheme promised investors a return of 1,000%, 'a cheat and a Quack', and his remarks about W. T. Brande were similarly caustic (*see* below, pp. 132–3). Davy refused to patent the safety lamp, and his services in the manufacture of gunpowder with J. G. Children, Davy told the public, were gratis.

[205] ibid., p. 138. With few exceptions it was only after he achieved national stature that he began to argue publicly for the value of science for its own sake. *See DW*, VII, 1–112, *passim*.

[206] Griggs, *Coleridge Letters*, II, 1042; dated 26 January 1804.

[207] Quoted by Siegfried in 'The Chemical Philosophy of Humphry Davy', p. 201.

3
Industry and Empire: competition for the control of science

In setting the Royal Institution along the path of agricultural improvement, the landowners had taken the entrepreneurial spirit of the Industrial Revolution and applied it to their own concerns. Given this fact, we would expect the representatives of industrial and commercial classes among the Proprietors of the Institution to be in an uneasy relationship with the landed interest, for although the ideology of science there was not of aristocratic origin, the RI was unequivocally in the hands of the landowning class. This was apparent from the outset. Halévy noted that the original proposal 'was not received favourably by the manufacturers'.

> Possibly [he continued] their natural selfishness and distrust took alarm at [Rumford's] request for a supply of models for exhibition. Would not such an exhibition betray their secrets to competitors? The members of the aristocracy, on the other hand, subscribed liberally. Less secretive than the manufacturers, they were no less in need of machinery for the exploitation of their estates.[1]

Very few industrialists joined the RI, for they correctly perceived it to be an organization run by the landed nobility, who had much to take from them and little to offer in return. 'Pure' industrialists, i.e. those without large sources of hereditary wealth, were poorly represented in the Proprietorship—none among the first fifty-seven men, and less than 6% by 1803.[2] During the first decade only one manufacturer, Alexander Blair, rose to the governorship of the Institution.

The core group at the RI was aware from the first of the opposition the Institution would provoke. Two days prior to the first meeting of 7 March 1799, Joseph Banks undertook a bit of personal diplomacy to avert this negative reaction, inviting Matthew Boulton and James Watt to dine at his house, along with Lord Morton (slated to be a Manager) and Count Rumford.[3]

[1] Elie Halévy, *History of the English People in the Nineteenth Century*. Vol. I: *England in 1815*, trans. E. I. Watkin and D. A. Barker (2nd ed.; London: Ernest Benn, 1961 reprint), p. 566.
[2] The first fifty-seven did contain one spokesman for the manufacturing interest, James Wortley, but he was not a manufacturer himself.
[3] Tew MSS, Boulton's diary for 1799, entry for March 5th. This collection, formerly in the Birmingham Assay Office, is now called the 'Matthew Boulton Papers' and is on loan to the Birmingham Public Libraries from the Matthew Boulton Trust. I shall, however, continue to use the

The attempt to win Watt's and Boulton's patronage failed, however; neither of them joined. Eleven months later the Earl of Aylesford wrote to Boulton that the 'inventions of Count Rumford, may be in time applied, to other uses, than domestic economy',[4] but as Boulton was not in Birmingham to receive the letter it was opened by his secretary, Mrs Charlotte Matthews, who notified Boulton that she was sending the RI ten guineas on his behalf.[5] Within the next two weeks Boulton had apparently visited the RI and was pleased with what he saw, and wrote to his son Matthew Robinson Boulton of the Institution's progress. 'The King,' he added, 'has taken an active & interesting part in it: as well as the Nobility Male & Female it seems to be going on wth great spirit & Vigour.'[6] Robinson, however, was much more astute than his father regarding the dangers of the RI to the manufacturing community. The Institution's Repository, he wrote to Matthew,

> will not in all probability be much relished by the British Manufacturer—Instead of diffusing general knowledge it will strike him as an institution for diffusing the benefits of his capital & industry . . ... It may be a very pleasant amusement for the nobility & other idle loungers who have never added an iota to the purse of the nation by the sweat of their brow, to diffuse the inventions & advantages acquired by the perseverance & painful study of groveling mechanics; but how will it be relished by the inventor himself . . ... An Institution for diffusing general knowledge & science may be useful, but if the manufacturers find that it is intended to be made a vehicle for disclosing the particular arts & machinery employed by them, their opposition to it will be found equally powerful as the support of the patrons, tho' composed of such exalted personnages [sic] . . ..[7]

This antagonism was not directed at the RI alone. The Society for the Encouragement of Arts and Manufactures was forced to close its model room in 1800 as a result of industrial pressure.[8] Rumford had asked the Boulton-Watt firm for a working steam engine so that workmen could make drawings of it, at the very time that the Boulton-Watt patent on the engine was due to expire.[9] Robinson's letter, his father wrote to him, 'made a strong impression upon my mind . . .'.

> It is a subject [he continued] I never reflected upon before & when M[rs] Matthews wrote me that she thought I should be a subscriber . . . I did not object but consented; since w[ch] I have seen Count Rumford at S[r] Jos[eph] Banks & likewise at the Institute in Albermall [sic] Street where he took an opportunity of Shewing me all the appartments [sic] & one for a Steam Engine which he says he must have . . . I

traditional citation of 'Tew MSS'.

On the following narrative *see also* W. J. Sparrow, *Knight of the White Eagle* (London: Hutchinson, 1964), pp. 127–30.

[4] Tew MSS, Box A, Item 244, dated 11 February 1800.

[5] ibid., note made on the letter by Mrs Matthews.

[6] Tew MSS, Box labelled 'Boulton, M. R., to Boulton, M. 1799–1842', Item 65, dated 24 February 1800.

[7] ibid., Item 67, dated 28 February 1800.

[8] *The Commercial and Agricultural Magazine*, II (1800), 450.

[9] Tew MSS, Box 'Boulton, M. R., to Boulton, M. 1794–1842', Item 67.

now from your letter see the thing in another light & think if a few such letters were published it would fall . . ..[10]

Such letters were apparently circulated. The Model Room, whose first items were largely related to the allotment programme of the maintenance of the rural poor, now became inactive. An inventory made at the RI on 23 August 1803 does not list many models, and a report of 31 December 1803 notes that 'the adequate supply of useful Models still continues a desideratum in our Establishment . . .'.[11] The idea of a model room was not abandoned, but, from 1804, the room began to serve as a repository for agricultural implements. Models of machines for raising water, one of a self-closing gate (for cottages), and of grinding mills began to fill the collection.[12] Lord Winchilsea also donated farm equipment, and the Visitors' report for 1804 recorded 'very considerable improvements . . . in the Libraries, the Laboratory, [and] the Model Room . . .'.[13] As we saw, John Hippisley's letter to Lord Hardwicke reminded the latter that the Repository contained (as of 1820) a number of models of agricultural machinery (above, p. 417). Boulton's activities, therefore, did not force the Model Room to close but rather to confine itself to agricultural instead of explicitly industrial machinery. If the landed aristocracy wanted to learn about manufacturing, they would have to do so at their own expense.

The case of the commercial interest at the RI is much more ambiguous, for unlike the manufacturers, they had nothing to lose. Furthermore, the life style of the nabobs, or returned civil servants of the East India Company, emulated that of the aristocracy as its ideal. As a result, merchants were naturally eager to join a scientific organization run by the representatives of polite society; and yet they were also attuned to a more entrepreneurial ideology of science which was also present at the RI, and which they had in addition derived from their colonial experience. The role of the 'colonial interest' at the RI is thus not as easy to describe as that of the manufacturing classes, and warrants a fuller discussion.

The penetration of English seapower into the Asian subcontinent dates from the late sixteenth and early seventeenth centuries, but it was only during the eighteenth century that the East India Company changed from a commercial enterprise to a territorial one. The victories of Lord Clive in particular made it the military arm of the English government in India, and the ouster of

[10] ibid., Item 68, 3 March [1800]. Boulton immediately changed his mind about becoming a Proprietor, although his name appears on the list for 1 May 1800. On 17 December 1801 he was sent a form that he had been elected a Proprietor, but the form is predated 31 March 1800 (Tew MSS, Box R2, Item 201).

[11] RI Archives, Box File XIV, Folder 130, pp. 12–16, and *The Annual Report of the Visitors, on the Accounts of the Institution to the 31st of December, 1803, and on the Progress and State of the Institution*, Box File XIV, Folder 128a.

[12] The Ambulatory at the RI contains a model of a water-raising machine and one of a self-closing gate. Mr W. J. Green, who worked at the RI between 1900 and 1950, and who was assistant to James Dewar, informs me that there had been models of grinding mills which were later thrown out.

[13] RI Archives, *The Annual Report of the Visitors on the Progress and Present State of the Institution, 22 April 1805*, Box File XIV, Folder 128a.

French and Dutch forces gave Britain virtual monopolies on saltpetre, opium, salt, cotton, raw silk, calicoes and indigo. In February and March of 1799 the EIC inflicted a critical defeat on the French and annexed some of India's richest areas. By 1812 the Company owned 115 ships, governed (in India alone) 380,000 square miles and sixty million people, and maintained a 150,000-man army.[14] With its military victories the Company began to collect revenue in the form of natural resources: £8 millions' worth in 1797, for example, and nearly twice that much per year by 1805. Legitimate trade did continue, however, with the Chinese, to whom English and Indian products were sold for tea—£55 millions' worth between 1793 and 1800.[15]

Strictly speaking, the EIC made no profits; rather, it gave its civil servants large salaries and lucrative opportunities. To obtain such employment it was necessary to find favour with the Court of Directors, a group of wealthy bankers and financiers. Men with sons to provide for commonly bought Writerships (the lowest rank of civil servant) from the Directors, so that although a Director's yearly salary was £300, the position was actually worth about £10,000 per annum.[16] Other factions involved in this imperial venture included the Board of Control, which supposedly exercised power over the Court of Directors; the 'shipping interest', a vast group of sub-interests involved in ship-owning and building, and/or private trade; bankers and businessmen whose financial dealing consisted primarily in subsidizing overseas investment; and nabobs, or returned civil servants, who frequently remained on the Company payroll and often represented the EIC in Parliament.

From this picture it is clear that thousands of Englishmen formed what can loosely be called a 'colonial interest', men who derived all or most of their income from England's colonial involvement. This group was heavily represented in the RI Proprietorship: nearly 37% of the first fifty-seven men were associated with the colonial interest. By 1803 this group was only slightly smaller than the landed interest, and this equilibrium was maintained throughout the first decade.[17] Before we can understand the nature of the relationship between these groups at the RI, however, we must first formulate some conception of the relations between landed and mercantile interests in general. T. S. Ashton argued for the fluidity of class lines: 'There was no sharp distinction between those who derived their incomes from rents and those who lived on the gains of commerce or industry'.[18] Just as some of the landed gentry

---

[14] C. N. Parkinson, *Trade in the Eastern Seas 1793–1813* (Cambridge University Press, 1937); C. H. Phillips, *The East India Company 1784–1834* (Manchester: University Press, 1940; reprinted, with minor corrections, 1961); and Holden Furber, *John Company at Work* (Cambridge: Harvard University Press, 1948).

[15] Parkinson, *Eastern Seas*, pp. 59–60, 80–1, and 96.

[16] ibid., p. 13. 'India House,' says Phillips, 'became a synonym for corruption and faction' (*EIC*, p. 23). Directors also made additional income from private investments over which they exercised significant control.

[17] There are problems in distinguishing some members of the EIC from the landed interest (*see* below). Eight of the first fifty-seven Proprietors, for example, belonged to both categories.

[18] T. S. Ashton, *An Economic History of England: The 18th Century* (London: Methuen, 1961), p. 21.

became interested in transport, or put their younger sons into trade, so did merchants and bankers seek to marry into the gentry or become country squires. The word 'merchant' was thus ambiguous. Lewis Namier has also pointed out that the younger sons of the gentry, and even of the peerage, often went into trade without loss of social status; and conversely, many successful merchants acquired country houses and invested in agriculture.[19] This was especially true of the nabobs, who customarily set themselves up as country gentlemen with a view to obtaining a seat in Parliament. About 6% of the M.P.'s who sat in the House of Commons over the period of 1734–1832 were nabobs or belonged to the colonial interest, and another 3% were associated with the West Indian trade. Thus despite the control of Parliament by the traditionally privileged class, 'there is considerable evidence of interpenetration between the aristocracy and men of business'.[20]

But the two classes were hardly the same, and the attitude of the landed interest toward the merchants was not one that bespoke social equality. The merchants in the eighteenth century, who came mostly from the retail trade, were 'an inferior group socially'.[21] They formed 'a motley race', not a true class like the landed gentry. There was no merchant aristocracy, despite the existence of wealthy merchants and bankers. Thomas Gisborne wrote of

> the aristocratic prejudices and the envious contempt of neighbouring peers and country gentlemen, proud of their rank and ancient family, who even in these days occasionally disgrace themselves by looking down on the man raised by merit and industry from obscurity to eminence . . ..[22]

The attitude of the merchants, wrote Lucy Sutherland, was a natural reaction to this, and it

> expressed itself in some resentment against and suspicion of the aristocracy and landed classes, with their easy arrogance and assumption of superiority, and in a somewhat self-conscious and self-righteous pride in their bourgeois virtues and bourgeois traditions . . ... Politically, there is evidence of a feeling that they and their interests lay outside the framework of a political system dominated by and organized for the interests of the aristocracy and landed classes, and an irritation because they, with their stake in the prosperity of the country and their contribution to it, were apt to be in the position of outsiders having to bring pressure to bear on the political machine.[23]

[19] Lewis B. Namier, *England in the Age of the American Revolution* (2nd ed.; London: Macmillan, 1961), p. 8; Conrad Gill, *Merchants and Mariners of the 18th Century* (London: Edward Arnold, 1961), pp. 11 and 135.

[20] Gerrit P. Judd, *Members of Parliament 1734–1832* (New Haven: Yale University Press, 1955), pp. 57 and 64–7.

[21] W. E. Minchinton, 'The Merchants in England in the Eighteenth Century', *Explorations in Entrepreneurial History*, X (1957), 64 and 67.

[22] Thomas Gisborne, *An Enquiry into the Duties of Men in the Higher and Middle Classes of Society in Great Britain* (2 vols.; London: J. Davis, 1795), II, 397.

[23] Lucy Sutherland, 'The City of London in Eighteenth-Century Politics', in Richard Pares and A. J. P. Taylor, *Essays Presented to Sir Lewis Namier* (London: Macmillan, 1956), p. 60.

Conversely, from the standpoint of the improving landlord, there may have been some hostility over the question of the deployment of national resources. Arthur Young was staunchly anti-imperialist, and probably typical of the rest of the Board of Agriculture in regarding colonial investment as a dangerous alternative to investment in the land. The former he saw as greedy and irresponsible; the latter as patriotic and enlightened.[24]

The relations between the landed and commercial interests were thus a complicated affair, and these complications would naturally carry over to an institution like the RI, which contained substantial numbers of each group. The major problem is to distinguish between the two interests as they affected the governorship. The first election, for example, brought three men from the colonial interest into office—J. C. Hippisley, R. J. Sulivan, and Lord Teignmouth—but both Hippisley and Sulivan had agricultural interests as well, and it is difficult to decide which interest prevailed in situations that would dictate a choice. By 1803 the number of the colonial interest had risen to five governors (out of thirty-one), yet all of them had agricultural interests (Hippisley, Hobart, Hobhouse, Lewisham, and Sulivan). Insofar as the colonial interest can be distinguished from the landed classes, whether by economic orientation or social and psychological attitude, it is clear that this group had no genuine representation among the governorship of the RI. In 1803 landowners outweighed the East India faction by a ratio of 6:1. In this way, the situation at the RI duplicated the general political situation described by Lucy Sutherland (above). If, for example, the colonial interest wanted to direct science into commercial rather than agricultural channels, it would simply not be able to succeed in doing so.

Much of the merchant interest in science lay in its ability to provide some osmosis between class lines. Nabobs often took up science as a pleasant pastime, joining scientific societies when possible.[25] Yet the interest went deeper than this, for science was regarded by the EIC as an integral part of British hegemony in Asia. Trade, as well as outright plunder, revolved around the plantation industries—indigo, hemp, tea, sugar (especially in the West Indies), rice, tobacco, spices, opium, cotton and other products. Agriculturalists, botanists, and physicians serving in the Far East were all enlisted in the attempt to introduce such crops in different localities, or improve the yields of existing plantations. Scientists at home also became involved in the cause, men like Sir Joseph Banks and the Earl of Dundonald.[26] As one historian has written:

[24] John Gazley, *The Life of Arthur Young* (Philadelphia: American Philosophical Society, 1973), pp. 82, 160–1, 163, and 268; Claudio Veliz, *Arthur Young and the English Landed Interest 1784–1813* (Ph.D. Dissertation, University of London, 1959), p. 278.

[25] James M. Holzman, *The Nabobs in England* (New York: 1926). *See also* the discussion by William Marsden, an original Proprietor of the RI, in *A Brief Memoir of the Life and Writings of the Late William Marsden* (London: J. L. Cox and Sons, 1838), pp. 25–6.

[26] Banks' role can be ascertained by consulting the abstracts of his correspondence with the EIC in Warren R. Dawson (ed.), *The Banks Letters* (London: British Museum, 1958). A letter from the Company secretary, William Ramsay, to the Earl of Dundonald, 16 June 1803, informing him that he will be paid for the experiments he conducted, is preserved at the IOL, Home

The Directors of the Company were renowned for their interest in profits and dvidends; but in the field of natural history, self-interest and scholarship often coincided. Self-interest, indeed, gradually led the Company to set up scientific institutions, make expert appointments, sponsor expeditions and subsidize publications. A knowledge of botany was found necessary for experimental horticulture and the exploitation of forests. It also assisted in medical and technical research . . .. Able Company servants, most of them doctors or engineers, helped to carry on such investigations; and . . . their work was encouraged by the Company because it might ultimately have solid financial results . . ..[27]

Nor was this use of science confined to natural history. The manufacture of gunpowder, for example, became a matter of chemical analysis in wartime. Normally, it was a simple affair: saltpetre was produced in the villages of Bengal, collected by the Company, sent to England and there blended with sulphur and charcoal.[28] In the last few years of the eighteenth century, however, 'chemical discoveries . . . enabled the French to dispense with a foreign supply . . .',[29] giving the EIC (if it needed it) a dramatic illustration of the military and economic importance of chemistry. The Company turned its attention to improving the quality of gunpowder, which its experts believed was strictly a question of chemistry. The Company also relied on chemistry for the assaying of ores in its mining operations, and for the manufacture of dyes. Finally, chemistry was seen as important for the investment in leather, a substance much in demand for harnesses and saddlery in India.[30]

The ideology of science that emerged from such a dependence (or perceived dependence) on the subject was, naturally enough, similar to that held by the landowning class, and the faith of the EIC servants in science was understandably strong. This is best seen in a document such as Lord Macartney's journal of his embassy to China in 1793, which was undertaken to open up that country as an official British market. The embassy, which included four future RI Proprietors, was to gather all possible information on the manufactures and commerce of China.[31] The original instructions for the mission were sent to Macartney by another future Proprietor, Francis Baring, the Chairman of the Court of Directors.[32] Macartney was also ordered by the Home Secretary to impress the Chinese 'by demonstrating England's scientific knowledge and

Correspondence, Miscellaneous Letters Sent, E/1/239, p. 141.

[27] Mildred Archer, 'India and Natural History; the Role of the East India Company 1785–1858', *History Today*, IX (1959), 736.

[28] Parkinson, *Eastern Seas*, p. 78.

[29] ibid., pp. 83 and 85.

[30] One example was the discovery, in 1803, of 'a vegetable preparation for preserving Leather', by a Company chemist, which the Directors immediately ordered to be put into use (IOL, Correspondence with India, Bombay Despatches, E/4/1019, p. 37, 13 January 1804).

[31] Earl H. Pritchard, 'The Instructions of the East India Company to Lord Macartney on His Embassy to China and his Reports to the Company, 1792–4', *Journal of the Royal Asiatic Society of Great Britain and Ireland*, 1938, pp. 212ff.; and J. L. Cranmer-Byng (ed.), *An Embassy to China; Being the journal kept by Lord Macartney during his embassy to the Emperor Ch'ienlung 1793–1794* (London: Longmans, 1962), p. 3. The future Proprietors included Sir George L. Staunton (a close friend of Joseph Banks), the secretary to the embassy; his son, George T. Staunton; Francis Baring's son John; and Dr William Scott, the ship surgeon.

[32] Pritchard, 'Instructions of the EIC', p. 203.

technical achievements'.[33] Macartney's first report shows that to him, science would be of major importance in opening up China to trade. Noting that thin leaves of tin are widely used for religious purposes, and that Malay tin is especially malleable, he adds:

[T]his accidental quality must probably be derived from some variety in the process of the reduction of the ore (such as using fuel perfectly freed from the smallest mixture of any particles of sulphur), which the advanced state of Chemistry in England might enable the artists there to find out and imitate, as a means of extending very much the sale of that article in China.[34]

And in his discussion of Chinese porcelain he remarks:

[T]hose specimens which I shall forward to Sir Joseph Banks with the view of having them compared under the eye of Chemists and skilful artists with the materials used in England for the same purpose, may afford an opportunity of judging if any improvement yet remains to be made in our manufactures of the same kind.[35]

In Lord Macartney we have the EIC's conception of science plainly stated. The chemist is not a theoretician, or even an experimenter, but a technician, an 'artist' at the service of the entrepreneur or the corporation, by means of which England would extend her commerce and improve her manufactures.

The EIC was also eager to establish scientific education for its employees, an effort that involved other future Proprietors of the Royal Institution. John Shore (later Lord Teignmouth), the Governor-General of India, sponsored the Macartney embassy's 'mechanic' (i.e. scientist) in giving 'a course of lectures in experimental philosophy, which at once established his fame and independence in Bengal'.[36] These lectures, given by a Dr Dinwiddie, who was also patronized by Lord Hobart (later a governor of the RI), were so popular with the Britons there that the Board of Trade had Shore appoint him for consultation 'in several parts of chemistry, mechanics, and natural philosophy, which have relations to the affairs under our charge ...'—i.e. bleaching, sugar and silk production, gunpowder manufacture, refining of native gold and silver, etc.[37] Dinwiddie went on to become Professor of Mathematics, Natural Philosophy and Chemistry at the newly founded (1799) College of Fort William, a college that John Shore and a second RI Proprietor, David Scott (a Chairman of the

[33] Cranmer-Byng, *Embassy to China*, p. 30.

[34] Pritchard, 'Instructions of the EIC', p. 385.

[35] ibid., pp. 391–2.

[36] W. J. Proudfoot, *Biographical Memoir of James Dinwiddie* (Liverpool: Edward Howell, 1868), p. 98.

[37] ibid., pp. 98–9. *See also* Cranmer-Byng, *Embassy to China*, pp. 310–11, and Dinwiddie's *Syllabus of a Course of Lectures on Experimental Philosophy* (London: A. Grant, 1789). James Dinwiddie (1746–1815) was one of the itinerant scientists typical of this period. He remained at Fort William College until 1805, when he returned to England and joined London scientific society, frequently attending lectures at the RI.

See also H. J. C. Larwood, 'Science in India before 1850', *British Journal of Educational Studies*, VII (1958), 37.

Court of Directors), did much to promote.[38] In 1805, when Haileybury College in Hertfordshire was established to provide three years of compulsory education (including science) for prospective EIC servants, its cornerstone was laid by Francis Baring and Edward Parry, the latter an original Proprietor who was on the Select Committee of the SBCP to confer with Rumford.[39]

There was, then, not only a significant overlap between the RI and the EIC in terms of individuals, but also in terms of ideology: the use of science as an instrument of economic policy. It was over this issue that tensions first arose. The Company interest in scientific education that began at the very end of the century had early repercussions in the plans of the Royal Institution. Shortly after Thomas Webster proposed his school for mechanics, the Prospectus of the RI was printed under the direction of R. J. Sulivan and J. C. Hippisley.[40] It was never circulated publicly, and the only extant copy belonged to the Earl Spencer.[41] It reveals that the authors had combined Webster's idea of a trade school with the very timely notion of having a training academy for EIC civil servants. Fort William College was being threatened by some of the Directors, due to their opposition to Governor-General Wellesley, and they wanted to abolish it and establish a domestic academy in its place, 'where mathematics, physics, and the elements of other science could be taught. . . .'.[42] We can thus imagine the interest of the colonial faction at the RI over the concluding paragraphs of the Sulivan-Hippisley prospectus:

> BESIDES the courses of public lectures it is proposed to admit pupils into all or any of the departments of research which constitute the objects of this establishment. The instructions to be given to the *pupils of the Institution* will be such as will constitute a complete course of scientific education, theoretical as well as practical,

[38] C. H. Phillips (ed.), *The Correspondence of David Scott, Director and Chairman of the East India Company, Relating to Indian Affairs 1787–1805* (2 vols.; London: Royal Historical Society, 1951), II, 304, 306, 313–14, 316, 358, 400, 427, and 431. Sir Elijah Impey, a judge in Calcutta and RI Proprietor, was also consulted on the plans for the college.

[39] The following publications contain information on Fort William and Haileybury: Phillips, *EIC*, pp. 125–30; Phillips, *Scott Correspondence, passim*; G. S. A. Ranking, 'History of the College of Fort William', *Bengal Past and Present*, VII (1911), 1–29, XXI (1920), 160–200, and XXIV (1922), 112–38; F. C. Danvers *et al.*, *Memorials of Old Haileybury College* (London: Archibald Constable, 1894); Thomas Roebuck, *The Annals of the College of Fort William* (Calcutta: Philip Pereira, 1819); Prakash Chandra, 'The Establishment of the Fort William College', *The Calcutta Review*, LI, Third Series (1934), 160–71. The original plans for Fort William College included professorships in geography, mathematics, natural history, botany, chemistry, and astronomy.

[40] *MM*, I, 75 (23 December 1799). Sulivan and Hippisley, with Banks' help, had also drawn up the bylaws at an earlier date, with Rumford only serving to get in the way (*see* letter from Hippisley to Banks, 13 March 1800, BM Add. MS. 33980, ff. 223–4, and RS Archives, *The Diary of Sir Charles Blagden*, III, 10 March 1800).

[41] Purchased by the John Rylands Library, Manchester, in 1892; *see Sketch of a Prospectus of the Royal Institution of Great Britain, incorporated by charter, M,DCC,XCIX*, the John Rylands University Library of Manchester, Press Mark 13711.6. The same information is obtainable from a copy of the Prospectus for 1800, the final version of this sketch, in the RI Archives, Box 14 (old numbering system), Folder 13, on which Bence Jones copied out all of Lord Spencer's notations, as well as the supplementary section mentioned below.

[42] Chandra, 'Establishment of Fort William College', p. 165. Haileybury was ultimately chosen to replace Fort William.

but variable according to the particular designation of each respective individual. The conciseness of an introductory address forbids any ample examination of the particulars of this most useful body of instruction. The mathematical sciences, as applied to the surveys of particular districts, or the geography of countries and kingdoms; astronomy, navigation; the construction and use of mathematical and other instruments of extended application; the theory, formation, and use of engines and machines of all kinds, dynamic, hydraulic, and pneumatic, for naval, military, or civil use; the natural history of the animal, vegetable, and mineral kingdoms; chemistry, geology, mining; the processes and numerous improvements of manufactories:—These are a few among the subjects which, in the detail and due arrangement, will form a system of useful education never yet offered to the Public; a system calculated to open new scenes of rational enjoyment to men of independent fortune, and eminently suited to give increased value to the treasures of nature and art which lie neglected around us. Young gentlemen intended for civil and military service either at home or abroad, as well as those who are destined to become manufacturers, engineers, or commercial men, will here find those means of instruction which have long been sought in vain in the more limited establishments of private life. Who can point out the consequences when men so educated shall quit their country for its general service, or for the government and regulation of our immense territorial and commercial appendages? The extensive continent of India, fruitful in every product, the Isle of Ceylon, the Moluccas, and other Indian islands abundant in minerals and other most valuable articles, may be considered as yet unexplored with regard to the advantages they are capable of affording under the direction of men so qualified.

It will not then be necessary to send out workmen of very inferior qualifications to take charge of establishments requiring mechanical or chemical knowledge. Our West Indian colonies offer desiderata equally valuable: and the rich harvest of acquisition in the arts, in trade, in local establishments, in the exploring and improving our mines, our agriculture, and every other department of public industry at home, has already, in part, begun to shew itself, and must increase with a rapidity proportioned to the growth and adoption of the principles of useful science.

This very desirable course of education will be no additional expence [*sic*] to the Institution; but on the contrary, will add to its funds. For many and obvious reasons it is proposed that the pupils shall pay an adequate sum for the expences of their instruction, which, though much less than could afford the same advantages by any more confined system, will nevertheless be more than sufficient to defray all charges, besides affording a liberal encouragement to the professors.

A particular Prospectus of this department will shortly be given. In the mean time it may be proper to observe, that the moral conduct and regularity of behaviour of the pupils, more especially those resident in the house of the Institution, will be under the particular guidance of the principal professor, subject to the inspection of the managers and visitors; and that this essential particular, as well as their progress in the sciences, must, from the local situation of the Institution, be almost constantly under the eye of their parents and friends.[43]

As the text reveals, the goal for which science is intended reflects unequivocally the ideology of science prevalent at the time, and certainly mirrors the concept of scientific education held by the EIC. The references to 'young gentlemen', 'pupils', and 'parents' strongly suggest that the students enrolled in this programme would not be rural labourers or lower-class mechanics, as Webster had wished, but the sons of the Proprietors themselves. Since service

---

[43] *Sketch of a Prospectus*, pp. 15–16.

in the EIC was purely a question of nepotism, it is clear that the Company servants in the RI Proprietorship, believing that science would enable future governors and Company employees to exploit India more efficiently, wanted those future rulers (i.e. their sons) to obtain a scientific education. Finally, lest there be any doubt as to the role of the EIC in this matter, Lord Spencer made a note in the margin worth quoting: 'This supplementary part has been abandoned, altho' in the opinion of many gentlemen connected with the *East & West Indies* especially, it was Thought highly useful & would have experienced great patronage'.[44]

Thus the plan for the EIC school was dropped, and Webster had his way. It was the landed, and not the colonial interest, that supplied the school with pupils. Whether or not this became a point of serious conflict between the two groups, and to what extent overlapping interests mitigated such a conflict, is not clear. But the matter of the school suggests that an important issue had been raised: whom was science going to serve at the RI? At this early date in its history, science for commercial use had been suggested by one group and rejected by another, presumably the one in power.

Davy's investigation of tanning was the next issue that cut across class lines, although the evidence raises more questions than it answers. Davy, it will be remembered, had discovered that catechu, an extract of the mimosa native to India, contained large amounts of tannin. The governors were interested in an oak bark or leather substitute, and planned to arrange for importation with the EIC. This they did without any loss of time. On 2 February 1802 a letter from Sir Joseph Banks was read before the Court of Directors, asking them to consider the importation of catechu for its use as a substitute for oak bark in tanning.[45] The Court referred the matter to committee for consideration; but in the interim the secretary sent letters to the Governors-General in Council at Bengal and Bombay in an effort to block any possible mass purchasing by private individuals, in the event that the substance proved to be a lucrative investment:

> It having been discovered that the Substance commonly called Terra Japonica, or Cotch [*sic*] . . . may be used instead of Bark in tanning Leather, &, according to some Experiments which have been made, with more advantage, as it will tan Leather in a few days, whereas Bark requires from six to thirteen Months, I am directed to notice the circumstance to you, that the Governor General in Council may prevent any unfair Speculations which, may be attempted to be made in this article.
>
> Should it hereafter appear that any Public Benefit would be derived from the importation of it into Europe by the Company, the Court will give Such directions as to them may appear proper—[46]

Ten days later, the Court sent a dispatch to the presidencies of Bengal, Bombay, and Madras, as follows:

[44] ibid., p. 15.

[45] IOL, Minutes of the Court of Directors, B/134, p. 1013.

[46] IOL, Correspondence with India, Bengal Despatches, E/4/652, pp. 128–9; dated 2 February 1802, the same day Banks' letter was read to the Court. *See also* Bombay Despatches, E/4/1017, p. 213.

There being reason to conclude from the issue of some experiments lately made by an eminent Chymist of this Country that the article known in various parts of India by the name of Cutch or Terra Japonica ... may be found to possess qualities that will render it useful in the tanning of Leather as a substitute for Oak Bark provided it can be afforded at a suitable rate of price, We direct that as soon as possible after the recei[p]t of these advices you cause the needful enquiries to be made as to the quantities of the article that can be procured & rates of price at which it can be afforded in case a demand should arise for the same in large quantities. It would also be useful if you obtain an account of the process by which it is fabricated, the places where it is procurable in the greatest quantities, & in the highest state of perfection.

This information accompanied with Samples of the article sufficient to give it a fair trial in a practical experiment, say One or Two Tons weight of each sort we desire may be transmitted as soon as possible. We conclude that D.ʳ Roxburgh at Bengal D.ʳ Berry at the Coast D.ʳ Scott at Bombay M.ʳ Campbell at Bencoolen & The Botanist at Prince of Wales's Island, will be ready to afford you material assistance in your enquiries on an object that may probably be of importance in a Commercial point of view to both Countries.[47]

At this stage, then, it is apparent that Davy's discovery had excited the imagination of the EIC, which saw in the importation of catechu a lucrative item of trade. Dr Roxburgh sent the Company a comprehensive report on the various types of catechu available in India and the tannin content of each, and suggested that the extract could be 'transported to England at a small expense, opening a new source of industry, which cannot fail to become interesting to all who are concerned'.[48] Writing from Bombay, Dr Scott also furnished complete details of his examination of catechu, informing the Court that 'catechu grows in the greatest abundance over all that [v]ast tract of Mountains which divide Hindoostan from North to South and would afford Materials for immense manufactures,' and he added that the natives were aware of its tanning powers.[49] Reports were also submitted by other botanists, Mr Campbell, Mr Hunter, Dr Berry, a Dr Heyne, and a man named Sir Paul Joddrell.[50] Dr

[47] IOL, Correspondence with India, Bengal Despatches, E/4/652, pp. 327–31. *See also* Madras Despatches, E/4/889, pp. 167–9 and 493; and Bombay Despatches, E/4/1017, pp. 405–9. The section quoted is specifically from the Bengal Despatches, and is dated the 2nd April, but the Madras Despatches indicate the date of the section quoted was actually February 12th. The following section was added later, and sent to Bengal on the 2nd April and to Bombay on the 26th March: 'Having ourselves procured some Information respecting the Terra Japonica since the preceding Paragraphs were written, we transmit you in the Packet a Paper which contains it'.
On William Roxburgh (1751–1815), Curator of the Royal Botanic Gardens, *see* Archer, 'India and Natural History'. Helenus Scott (1760–1820) was in the Company's medical service. The 'botanist' referred to is probably William Hunter (1755–1812), surgeon and orientalist.

[48] IOL, Bengal Public Consultations, Range 5, Vol. 36, No. 44 of 19 August 1802; the letter is dated 27 July 1802. *See also* Correspondence with India, Letters Received from Bengal, E/4/64, Commercial Letter of 12 January 1804.

[49] IOL, Bombay Commercial Proceedings, Range 414, Vol. 72, 24 December 1802, pp. 1324–30.

[50] Hunter's report is in the Bengal Public Consultations, Range 5, Vol. 38, Nos. 27–8 of 19 August 1802 (dated 4 November 1802). Berry, Heyne and Joddrell are mentioned in Correspondence with India, Letters Received from Madras, E/4/329, Commercial Letter of 20 October 1802, paragraphs 6–10, and E/4/331, Commercial Letter of 9 May 1803, paragraphs 46–8. Campbell's report is in Bengal Public Consultations, Range 5, Vol. 41, No. 18 of 6 January 1802. *See also* Correspondence with India, Letters Received from Bengal, E/4/63, Commercial Letter of 30 October 1802, and of 21 March 1803.

Berry's report, plus several specimens of catechu and a piece of leather tanned with the substance, were sent to Banks, 'with a Request that he will procure the Opinions of Chemists and Scientific Persons thereon as speedily as may be convenient'.[51] Banks complied with the request within two months' time, submitting a report on the leather tanned with catechu and the specimens of the extract, and 'on the probability of Cutch becoming an article of consumption in this country . . .'.[52] In August of 1804 the Court of Directors drafted the following dispatch to Bengal:

> The increasing scarcity of Oak-Bark renders the Importation of other Tanning Materials an object worthy of Consideration; and the Legislature having passed an Act, allowing for a further term of seven years the conditional Importation of Foreign Oak Bark.* We shall procure the Opinions of Chemical Gentlemen on the several Specimens of Cutch which we have received from our Presidencies; — and some important observations from the Right Hon[ora]^ble Sir Joseph Banks on that article will be found in the Packet.* Nevertheless, *as the subject is of importance to the landed Interest of these Kingdoms;* You will not, without our particular Instructions, provide a greater quantity of Cutch than may be sufficient for correctly ascertaining the qualities and prices at which it can be furnished.[53]

The subject was thus very much a live issue more than two years after Davy's work. It was still 'of importance to the landed Interest', particularly the governors of the Royal Institution, and therefore to the colonial interest as well, for it was apparent that importation of catechu would be a profitable commercial undertaking. Yet the extract was not imported. Writing to Madras, the Directors stated that a 'sufficient quantity of this article having been imported to make its properties generally Known: We do not propose for the present to continue it in our list of investments'.[54] John Burridge, a tanner who wrote to the Company regarding catechu years later, was told by the Court that the Directors had seen 'it right to leave the further prosecution of the subject for private import'.[55] The reply, however, is clearly unsatisfactory: it made no sense whatever to talk of leaving the matter 'for private import', as importation of catechu was illegal prior to 1808 and could have been made legal only through Company influence. The more likely explanation was that the Com-

[51] IOL, Minutes of the Court of Directors, B/138, pp. 989–90, 30 December 1803; and *see also* Home Correspondence, Miscellaneous Letters Sent, E/1/239, p. 366, 2 January 1804.

[52] IOL, Minutes of the Court of Directors, B/138, p. 1252, 29 February 1804. The Company responded to Banks' comments on Dr Berry's report by supplying him with more samples of catechu for further experiments.

[53] IOL, Correspondence with India, Bengal Despatches, E/4/656, pp. 517–18, dated 21 August 1804. The last sentence, which was to be inserted after the first or second asterisk (the text leaves this ambiguous), was crossed out in favour of the following: 'Nevertheless we do not wish for any larger assignment of the article from you than may be sufficient for full & complete experiments of its qualities, until you receive our further information & orders'. The italics (in the original version) are mine.

[54] IOL, Correspondence with India, Madras Despatches, E/4/894, p. 259, 10 April 1805.

[55] John Burridge, *The tanner's key to a new system of tanning sole leather, of the right use of oak bark, &c, stating several most important discoveries and documents, in addition to Sir Humphrey [sic] Davy's experiments and the late Sir Joseph Banks' reports to the Honourable East India Company* (London: Charles Frederick Cock, 1824), p. 76.

pany found it more profitable to use it in India than to import it to England. The native demand for catechu was great, as it was used in chewing tobacco, paints, dyes, house plaster, sedatives for horses, and ointments for wounds.[56] The Company had also set up tanneries at Kanpur (formerly Cawnpore) and may have used catechu there.[57] Finally, catechu was needed in Asia to adulterate tea. Tea was the Company's greatest source of revenue: between 1793 and 1800 the total volume of sales for Indian and Chinese goods came to £103 million, of which £55 was the sale price of tea (*see* above, p. 98). G. L. Staunton, the Macartney embassy's secretary, wrote that any hindrance to the tea trade 'might be considered as a national calamity'.[58] In England the major, or lower-class, market was captured by means of a cheaper adulterated tea, and catechu was one of the principal adulterants used.[59] It was thus far more important for tea than for leather, but as adulteration was theoretically illegal, a candid reply to Banks' inquiries or to Burridge's was not readily forthcoming.

What were the repercussions of this in the RI? A definitive answer cannot be given, but as in the case of the proposed EIC school, the issue raised was whom science was meant to serve. It was the landowners who requested Davy to study leather manufacture, and who had much to gain from an oak bark substitute. As the letter of August 1804 (above) indicates, the EIC was fully aware of the importance of catechu to the landed interest, and vetoed its importation because it served their own interests to do so. Although we can only speculate, it seems reasonable to assume that this deepened the issue of institutional control.

In 1804, the RI was presented with a proposal which could have mitigated some of this conflict, viz. the Mineralogical Collection, discussed briefly in Chapter 2. The landowners' interest in the project was due to the relevance of geology for mining, and Davy had already analysed slate from Lord Penrhyn's quarry and lead from a mine belonging to Sir John Sinclair.[60] The EIC interest in the subject was equally obvious, for a knowledge of mineralogy was seen as an aid to the exploitation of Asia's natural resources, and assaying had value for the refining of native ores. Although the proposal was adopted, it did not succeed in reconciling the two factions at the RI because it became clouded over by a much larger problem with which the Institution was not equipped to deal, namely science for public vs. private use. By 1805 the East India men at the RI had, for all intents and purposes, left to form their own establishment, so if the Mineralogical Collection was a compromise venture, it was too little and

---

[56] IOL, MSS. Eur. E. 11, James Kerr's 'Observations upon Natural History', section entitled 'A Description of a New Plant from which the Terra Japonica of the Shops is extracted', ff. 1–6.

[57] In 1801 Cawnpore became the chief British frontier station, and the manufacture of leather goods there became a large-scale industry.

[58] G. L. Staunton, *Historical Account of the Embassy to the Emperor of China* (London: John Stockdale, 1797), p. 15.

[59] Parkinson, *Eastern Seas*, p. 95.

[60] RASE Archives, *Rough Minute Book*, 1803–6, p. 63 (27 March 1804); A. and N. L. Clow, *The Chemical Revolution* (London: The Batchworth Press, 1952), p. 364.

too late. But the story is worth relating because it did raise the problem of institutional control once again, and in a more general way; and it was this issue that ultimately forced the changes of 1809–11.

On 10 April 1804 three men, Sir John St Aubyn, Sir Abraham Hume, and Charles F. Greville met

> to consider . . . the propriety & means of forming a Society for the Establishment of a National Collection of Mineralogy with a Lab[o]ratory for the advancement of Metallurgy.[61]

St Aubyn, a wealthy landowner, was a prominent figure in Cornish politics and a partner in the Cornish Copper Company;[62] Hume was a wealthy landowner and possibly a shipowner in the EIC;[63] and Greville had served as a Lord of Trade, a member of the Board of Trade and Plantations, and like St Aubyn and Hume was a collector of minerals.[64] Asserting 'that no Existing Establishment has provided for the adequate advancement of Mineralogy', and

> that Mineralogy requires a distinct Establishment in which the whole grounds of Mineralogy & Geology illustrated by Specimens and authenticated by Chymical Analysis can be invest[ig]ated,

they resolved to set up such an establishment along the lines of the Royal Institution.[65] When they communicated their intentions to the RI on 5 May 1804, the Managers decided to make the proposed organization a part of the Institution itself, and ordered Thomas Bernard and Charles Hatchett to confer with the three men.[66] The plan (*see* Chapter 2) was to raise £4,000 by subscription, in order to finance a mineralogical collection at the RI, and an additional laboratory, to be devoted to the assaying of metals and the advancement of metallurgy. The Earl of Morton, Lord Dundas, and Bernard conferred with the three men and prepared an address to the Proprietors and Subscribers, which was also sent to the ministers of the cabinet, the members of both houses of Parliament, the Directors of the EIC, the members of the Board of Agriculture, as well as 'the different Mineral Companies in this Kingdom . . .'.[67] The final proposal stated

[61] BM Add. MS. 42071, ff. 112–13.

[62] D. B. Barton, *A History of Copper Mining in Cornwall and Devon* (Truro: Truro Bookshop, 1961), p. 24. *See also* W. P. Courtney, 'Sir John St. Aubyn', *DNB*, L (1897), 121–2, and James Greig (ed.), *The Farington Diary* (8 vols.; London: Hutchinson, 1808–9), V, 278 ff..

[63] W. P. Courtney, 'Sir Abraham Hume', *DNB*, X (1908), 208–9; and John Mathison and Alexander W. Mason, *An East-India Register and Directory, for 1803* (London: Cox, Son, and Baylis, 1803), p. xxv. Several of Hume's family were in the EIC. He was also known as a collector of precious stones.

[64] Dawson, *Banks Letters*; and Alan Valentine (ed.), *The British Establishment 1760–1784* (2 vols.; Norman: University of Oklahoma Press, 1970), I, 391. Greville was also an Honorary Member of the Board of Agriculture.

[65] BM Add. MS. 42071, ff. 112–13.

[66] *MM*, III, 276 (7 May 1804); and letter to the three men from the RI Secretary, J. P. Auriol, BM Add. MS. 42071, f. 114.

[67] *MM*, III, 279 and 281 (17 May 1804).

that the mining concerns in this Kingdom are conducted by individuals with such advantages of capital, and with such a degree of speculative enterprise, as to exhibit those effects of combined capital and mechanical powers applied to them, which no other country in the world has hitherto been capable of producing . . ..[68]

It also directed itself to the East India Company:

But in the immense Territory, which now forms our East India Possessions, are to be found the most valuable mineral treasures that are known in this globe; and from the wisdom and liberality of the *East India Company*, great and effectual assistance may be hoped for in aid of the execution of a plan, by the adoption of which the intrinsic value of those treasures may be ascertained and brought into use.[69]

The proposal concluded on the note that this plan was part of the original goal of the RI to diffuse science. The Managers resolved to send this information to the Governor-General of India (Wellesley), as well as the Governors of Fort St George and Bombay, asking them for help in putting this proposal into action; and they appointed J. C. Hippisley, J. P. Auriol, and R. J. Sulivan (all nabobs) a committee of three to set up the Assay Office and Mineralogical Collection, and take steps 'for promoting the Circulation of the same in India'.[70] Their letter, which was also sent to the Governors of Ceylon and St Helena, as well as to Fort Marlborough and Canton, asked the EIC to permit subscriptions and minerals to be transmitted through its offices, and the Company agreed to cooperate.[71] Under Davy's direction, Charles Royce, Clerk of the Works, undertook to design a special room for the collection, the cost of which came to £186.[72]

The plan succeeded in attracting the kind of publicity the Managers wanted. In late 1804 and early 1805 the number of Annual Subscribers rose sharply, and Davy was soon receiving minerals from Proprietors in Lancashire and India.[73] In an effort to capitalize on their success, the Managers decided to throw Davy's services open to the public:

[T]he Laboratory shall be open for the analysis of such substances as the Managers or the Professor of Chemistry shall deem of Scientific or Public Importance . . . in case *any Person*, shall desire to have an Analysis of any Ore, Mineral, Soil, or other Substance within the British Dominions, which shall not be deemed of Scientific or Public Importance, the same shall be made at their Expense; the Sum to be fixed by the Managers, and not to exceed Ten Pounds for any one Analysis.[74]

---

[68] RI Archives, *Address to the Proprietors, Subscribers, and Others, respecting the proposed Mineralogical Collection and Office of Assay*, Box File XIV, Folder 146.

[69] ibid.

[70] *MM*, III, 289 and 293 (29 May and 4 June 1804).

[71] ibid., pp. 308–9 (25 June 1804), and IOL, Correspondence with India, Bengal Despatches, E/4/656, p. 581, and Madras Despatches E/4/1019, p. 483. *See also* letter to RI from Fort St George, 8 September 1805, in RI Archives, Box 15 (old numbering system), Folder 128.

[72] *MM*, III, 328 and 330 (10 and 17 September 1804), and IV, 4 (11 January 1805).

[73] ibid., III and IV, *passim*; and IV, 16 (28 January 1805).

[74] ibid., IV, 36–37 and 40. Italics mine.

As Hume, Greville, and St Aubyn conceived of it, the Assay Office and Mineralogical Collection would benefit industry, such as the Cornish Copper Company, and the EIC, whose wealth, they wrote, 'may be so greatly augmented'.[75] They believed that the plan would 'no doubt induce all the great Capitalists in mining, and all the great Landowners to contribute', but asked that subscriptions to the plan 'be open to such persons as we shall invite', among others.[76] Certainly, many people took advantage of the new arrangements, and Davy was so busy with the collection that by February of 1806 he was asked to work on it for the rest of the year instead of lecturing.[77] Few, however, contributed to its support, and indications are that of those who did, many were bankers and financiers. Of the £942 collected, £250 was supplied by Greville, St Aubyn, and Hume, and £100 by the famous banker Thomas Coutts.[78] Furthermore, the Minutes indicate that Greville, St Aubyn, and Hume were not even allowed access to the collection (!), since the Managers ruled in 1806 that if the three founders of the collection wished to use it, they would have to become Proprietors, a privilege which by then had risen to a cost of 150 guineas.[79] To some extent, this was unavoidable; the high cost of this privilege had to mean that Proprietors obtained something for their investment which the general public did not. From the point of view of the public, however, or of the three men who initiated the project, the governors had betrayed the whole programme. For the Managers had embraced the plan whole-heartedly, and had given it a good deal of public advertisement. When this attracted a large number of cash customers, in the form of Annual Subscribers, they threw Davy's services open to the public, available to 'any Person'. But for all the public service rhetoric, the Collection and Assay Office never served a wider public than the Proprietors themselves. Writing to Charles Greville in 1808, one of his friends congratulated him on his intentions with regard to the RI, adding, '[f]or the sake of science I wish the Splendid institution had been more come-at-able and that the world might have had more frequent opportunities of reaping the benefit of your exertions & munificence—'[80] When Greville died the following year, his brother Robert offered to sell his mineral collection to the RI, and Davy and Hatchett appraised it at £5,000.[81] Robert Greville finally sold the collection to the British Museum, and a letter he received from a friend on the subject says much about the image the RI had acquired for itself:

[75] RI Archives, 'Draft of the proposed Address to the Proprietors Subscribers & others, respecting the Mineralogical Collection & Laboratory of Assay', Box File XIV, Folder 146.

[76] *MM*, III, 297 (28 May 1804).

[77] ibid., IV, 149 (24 February 1806). On 24 March 1806 the collection was so large that it had to be moved to new quarters.

[78] ibid., III, 317 (9 July 1804), and IV, 192–5 (26 May 1806). The Minutes say that Coutts' name will be added to the list of bankers who have subscribed to the collection. Coutts had become a Proprietor in 1803.

[79] ibid., IV, 192–5 (26 May 1806). Their reply to the Managers, withdrawing from the plan, is in RI Archives, Box File XIV, Folder 146.

[80] Letter from Thomas Allan to Greville, postmarked 19 December 1808, BM Add. MS. 42071, ff. 155–6.

[81] BM Add. MS. 40716, f. 56, dated 27 December 1809.

. . . you know I never could Reconcile [myself] to the Idea of 5000£ for the Collection. It is very Clear M.ͬ Davy, & others of His Class wish to get It at a Bargain; & Make Their profit . . ..[82]

The Institution was not able to divest itself of the stigma of a private corporation run by and for the landowning class because, for the most part, that was what it was.

That institutional control, and the uses of science, was an issue at the RI is strongly suggested by the formation of a breakaway movement in 1805 by Sir Francis Baring, a Director of the East India Company. The London Institution, as it was called, pledged to bring 'Science and Commerce into contact', thereby adding to the commercial success of England.[83] It was, from its inception, an institution for merchants and bankers. The original founders and early elected governors of the LI included twelve RI Proprietors; within a few years twenty-three more RI Proprietors had swelled its ranks. Most of those that left the RI for the LI were associated with colonial, financial, or City interests; with one exception, they were unequivocally non-aristocratic and not of the great landowning class.[84] But apart from the changed constituency, and the projected reorientation of science toward commercial ends, the two institutions were virtually identical. The London Institution had committees of Managers and Visitors, elected for three-, two-, and one-year terms; three classes of subscribers, and a library, laboratory, and lectures.[85] The draftman's plans of the lecture theatre are very similar to the blueprints of the Theatre at the Royal Institution,[86] and the LI planned to have lectures on chemistry, mineralogy,

[82] Letter from J. Harpur to R. F. Greville, n.d., BM Add. MS. 40716, f. 107.

[83] R. Watson Frazer, *Notes on the History of the London Institution prepared mostly from the minutes* (London: Waterlow and Sons, 1905); *Plan and Bye-Laws of the London Institution, for the Advancement of Literature and the Diffusion of Useful Knowledge* (London: Phillips and Fardon, 1806); *The Charter, the Act of Parliament and the By-Laws, of the London Institution* (London: Waterlow and Sons, 1902); *An Historical Account of the London Institution* (London: reprinted from the introductory matter prefixed to the class catalogue, 1835). The original papers and minutes of the LI for 1805–17 are kept in the Guildhall Library, London. For a recent study of the LI *see* J. N. Hays, 'Science in the City: the London Institution, 1819–40', *British Journal for the History of Science*, VII (1974), 146–62.

[84] The twelve were J. J. Angerstein (who became a Vice-President of the LI), Francis Baring (who became President), Samuel Boddington, Richard Clark, Henry Hoare, William Manning, W. H. Pepys, John Rennie, Henry Thornton, Thomson Bonar, Benjamin Harrison, and William Saunders (the last three joined the RI after 1803; Bonar was a merchant and friend of Angerstein, Harrison a Deputy-Governor of the Hudson's Bay and South Sea Companies and Chairman of the Exchequer Loan Board, and Saunders a physician).
The twenty-three who followed were Nathaniel Atcheson, Robert Bingley, Thomas Boddington, Joseph Bradney, James Burton, Lord Carrington, Thomas Davy, Thomas Day, John Esdaile, Nathaniel Fenn, Joseph Hurlock, Johnson Lawson, John Lubbock, William Marsden, William Mount, John Pearson, Thomas Rowcroft, William Taylor, Garnett Terry, D. P. Watts, John Martineau, J. G. Ridout, and Samuel Scott (the last three joined the RI after 1803; Martineau was a partner in the Whitbread brewery, Ridout a doctor, and there is no information on Scott).
The one exception cited is Lord Carrington, who was a well-known banker before he became a peer and President of the Board of Agriculture.

[85] *The Charter . . . of the LI*, pp. 10–12; Frazer, *Notes on the History of the LI, passim*; Guildhall Library, LI Papers, MS. 3080, *passim*.

[86] BM Add. MS. 31323, 'London. *Miscellaneous Buildings*', Q² and R².

natural philosophy, and botany. There would also be a mineralogical collection, and a repository of models and machines in pneumatics, hydrostatics, hydraulics, mechanics, optics, electricity, and galvanism.[87] Finally, Baring asked Davy for his suggestions regarding the future objects of the London Institution, and the reply was enthusiastic. Davy urged Sir Francis to build a museum, or repository, 'furnished with the most important articles employed in Commerce, & in the useful Arts & Manufactures, in their raw stages & in their different states of preparation', as well as a laboratory.[88] Baring had also asked Davy about the wisdom of the LI acquiring the power of granting degrees in medicine—the type of middle-class concern in which the RI had shown no interest—and Davy again responded approvingly.

> I do not see that any but good consequences could ensue.—it would scarcely excite jealousy in the English Universities [!] & it would increase the utility & popularity of the medical lectures annually delivered in London.[89]

Davy concluded his letter by saying, 'I shall at any time be happy to attend your Commands, and I hope you will accept a general offer of my services in this business', a statement that may possibly suggest something beyond mere courtesy.[90] Davy never did lecture at the LI, but Brande later gave a talk there 'on the Connection between the Scientific and Commercial interests of a country, and especially those of Great Britain'.[91]

The career of the colonial interest at the RI was both brief and uninspiring. The pattern that emerges here is the larger one described by Lucy Sutherland (above, p. 79): merchants 'lay outside the framework of a political system dominated by and organized for the interests of the aristocracy and the landed classes . . .'. Their role was that of a pressure group rather than a controlling influence. Still, the story of the London Institution would seem to confirm the point already made, that due to the RI's imaginative use of Davy, science had taken on the appearance of a saleable commodity, and that the RI had become a model for how it might be deployed. The change from science-as-avocation to science-as-enterprise required that it sell itself, in its applied form, to sources of wealth; and thus the early history of the Institution, and the emergence of Davy as a new scientific type, proved to be crucial precedents for future scientific organization. Many institutions modelled themselves directly on the RI, or conceived of their laboratories in technological terms. At the very least, they emulated the ideology of science which the Royal Institution had come to represent. The only real difference between the RI and the numerous institutions that copied it was not purpose, but constituency. By appropriating

[87] Guildhall Library, LI Papers, MS. 3080.

[88] 'Fac-simile of a letter from Sir H. Davy to Sir F. Baring', 3 October 1805, in RI collection of *Miscellaneous Reports, &c.*, Series P, Vol. 2. A rough draft of this letter is in the Davy MSS, Vol. 22, notebook a, pp. 132–119 [*sic*].

[89] ibid.

[90] ibid. In the MS version of this letter Davy had written, after the word *and*, 'to offer my services in any way for the attainment of the object,' which he then crossed out.

[91] *Historical Account of the LI*, p. 27, and *QJS*, VII (1819), 205–22.

science to its own particular purposes, the RI provided an incentive for other groups to do the same.

If the London Institution is the best example of this, many others can be found. Thus Irish landowners decided to set up a scientific-agricultural institution of their own, and began their 'prospectus of the Cork Institution for the Application of Science to the Common Purposes of Life' with the sentence: 'The model of this establishment is to be found in the Royal Institution of Great Britain ...'.[92] After noting that a Reverend Thomas Hincks had already lectured on chemistry, the prospectus stated that the entire plan had been approved by the Duke of Bedford, then Lord Lieutenant of Ireland. In addition to a lecture room, the Cork Institution would have a laboratory, a library, a mineralogical collection, a 'store for the most approved implements of husbandry', a room 'for the use of the farming society or committee of Agriculture', and a botanical garden, 'the objects of which will be chiefly agricultural ...'.[93] The prospectus concluded by noting that the Dublin Society and Farming Society of Ireland wish to work together with the Cork Institution, and desire that similar societies be formed throughout Ireland.

> The interests of agriculture are so generally and justly regarded as of the first importance, that the managers will ... adopt whatever measures may appear most calculated for the advancement of it ....[94]

Many other organizations, while not explicitly agricultural, closely copied the set-up of the Royal Institution and its conception of science. Frederick Accum left the RI in 1803 to lecture at the new Surrey Institution, and other such places included the Philomathic Institution (1807), the General Institution (ca. 1805), the Russell Institution (1808), the Royal Institution of Cornwall, the Royal Manchester Institution (1823), and the Liverpool Royal Institution, 'clearly inspired by the London body ...'.[95] Thomas Kelly also claims that the '[i]nterest in applied science which had inspired the original Royal Institution was ... reflected in such bodies as the Scottish Society of Arts (1821), the Royal Cornwall Polytechnic Society (1833), and the Liverpool Polytechnic Society (1838)'.[96] These societies, like the provincial literary and philosophical societies that became numerous in the 1820s and 1830s, engaged in 'the delivery of Lectures, the formation of a Museum, the collection of a Library, and the establishment of a Laboratory fitted up with Apparatus'.[97] The Royal Institution thus served to stimulate provincial as well as metropolitan societies. As J. T. Merz has written, the RI 'has been repeated on a larger or smaller scale in many provincial societies, and notably in the colleges of Manchester, Bir-

[92] Reprinted in SBCP *Reports*, V, Appendix, pp. 119–22; dated 16 January 1807.
[93] ibid., pp. 120–1.
[94] ibid., p. 122.
[95] Thomas Kelly, *A History of Adult Education in Great Britain* (Liverpool: University Press, 1962). The General Institution is mentioned in the *Gentlemen's Magazine* for 1805; *see also* Frazer's *Historical Account of the LI*, p. 2.
[96] Kelly, *History of Adult Education*, p. 115.
[97] ibid., p. 113.

mingham, Liverpool, Newcastle, Leeds, Bristol, Nottingham, &c'.[98] In 1802 Thomas Bigge, a Vice-President of the Newcastle Literary and Philosophical Society, urged the creation of a lectureship in natural philosophy, to be called the New Institution:

> The late foundation in Albemarle-street, on a scale so magnificent and extensive, is a noble instance of the spirit of the times, and augurs well to [England's] best interests. But the Royal Institution of England ... is in one respect limited in its operation. All cannot frequent, for demonstration, the metropolis of their respective counties. Provincial establishments are therefore evidently necessary.[99]

It is interesting to note that the Bishop of Durham donated £100 to their effort, and that their first purchase was the philosophical apparatus that had belonged to the RI's first lecturer, Thomas Garnett.[100] During the first twenty-five years of the Newcastle Literary and Philosophical Society itself, its most important concerns were mining and agriculture.[101]

Thus the problem of who would control the direction of scientific activity was solved by the multiplication of institutions like the RI, but run by different groups. The problem of science for public vs. private use, however, was less easily resolved, and one could legitimately argue that, *mutatis mutandis*, it is still with us today. During the first decade the RI was ostensibly a public and national body, devoted to the 'application of Science to the common Purposes of Life'. In actual fact, it was a closed corporation, dominated by the landed aristocracy, providing service largely to its governors and at most to its Proprietors, and also to the Board of Agriculture. Virtually the same men were elected to the governorship year after year; Proprietorships were hereditary; and it was not by chance that the projects selected for scientific investigation were relevant to the concerns of the landowning class. Funding, after the initial investment, was supposed to come from the Annual Subscribers, but the exclusive nature of the RI discouraged their participation, and the Institution was saved only by large loans from the governors in 1803 and 1808. The horns of the dilemma were obvious: to use science for private interests necessitated continued private investment in the Institution; to put science at the service of the public and the nation meant rendering the Proprietorship superfluous. By 1809, it was clear that the structure of the RI would have to change if it were to survive.

In March of 1809, the Managers drew up an unusually candid report, in which they confronted the issue directly. Noting that the Institution had oc-

[98] John T. Merz, *A History of European Thought in the Nineteenth Century* (4 vols.; New York: Dover Publications, 1965 reprint), I, 249 n.

[99] Quoted in Robert S. Watson, *The History of the Literary and Philosophical Society of Newcastle-upon-Tyne (1793–1896)* (London: Walter Scott, 1897), p. 212.

[100] ibid., pp. 213 and 215–16. Garnett, in fact, was originally supposed to be the lecturer at the Newcastle Literary and Philosophical Society in 1799, but changed his plans when he received an offer from the RI.

[101] ibid., p. 151. The New Institution merged back into the Newcastle Literary and Philosophical Society in 1835.

casionally fallen into heavy debt, only to be saved by the willingness of the governors to contribute from their own pockets, it went on to say that if the support of those interested in science was to be obtained from the RI,

> something must be done to give it more the form of a Public Establishment, than of Private and Hereditary Property. It can hardly be expected that a general Interest should be excited for the Improvement of the Inheritance of a few Individuals . . ..
> [The Proprietors must] sacrifice . . . personal interest and advantage to erect a public, national, and permanent Establishment, devoted and dedicated to the cultivation of science, and to the promotion of every improvement in Agriculture, Manufactures, and the useful Arts of Life, that may be conducive to the happiness and prosperity of the British Empire.[102]

The first step made toward altering this state of affairs was a notice sent out to the various branches of the government on 10 April 1809.[103] Taking advantage of the recent decision of the Board of Admiralty to abolish the offices of Chemist and Metal Master, the Managers stated:

> [C]onvinced that the Chemical examination of Metals and other Substances, required by the Board of Admiralty, will be made more advantageously, and with less expense to the public, in the Laboratory of the Royal Institution, and under the direction of such a Scientific Committee of its Members as may be formed with the concurrence of the Lords Commissioners of the Admiralty, the Managers think it their duty to offer to them every assistance, as to Chemical Analysis and Experiments in Metallurgy, which is within the scope of their Establishment.
> . . . The investigations which the Royal Institution may offer, are applicable to many objects, apparently trivial in themselves, but really and essentially of the highest importance; such as the Analysis of Copper, Iron, and other Metals; of Drugs, Paints, Alcohol, Wines, Malt Liquors, and Water; the seasoning of Timber, the Tanning Substance, and Soils, Acids, Alkalis, and the variety of objects, in which the public welfare is eventually interested.
> They have therefore resolved, that a tender may be made to the Lords Commissioners of the Admiralty, of the aids and attention which they can supply, towards the examination of any Substances which may be sent by order of the Lords Commissioners to the Royal Institution, and in any Scientific experiments and investigations, which may be required by their Lordships . . ..
> The Managers being also of opinion that there may be many similar objects of original or collateral investigation, in other departments of [H]is Majesty's Government, particularly in the Military Department, in the Board of Ordnance, in the Offices of the Surveyors General of the King's Land Revenue, and of the Woods and Forests; in that of the Board of Agriculture, and of the Lord Warden of the Stanneries, &c., deem it proper that copies of this Resolution be presented to those Departments, offering the assistance of the Institution, with respect to any objects of Experiment, or Analysis within the scientific means of this Establishment.[104]

[102] RI Archives, *The Report of the Committee of Managers, upon the present state of the Institution*, 20 March 1809, Box 15 (old numbering system), Folder 33.

[103] This may be found in RI Archives, *Guard Book*, Vol. 1, and inserted between pp. 440 and 441 of *MM*, IV.

[104] ibid.

The replies to this offer of public service, which unlike the previous open-laboratory policy did not stipulate a fee for analyses, were generally encouraging, and must have made the Managers feel they were on the right track. The Surveyor General of the Land Revenue of the Crown, for example, wrote that he was responsible for the Navy's decision to abolish the offices of Chemist and Metal Master, but was now glad the RI had made its offer. It 'cannot,' he wrote,

> be doubted that the Royal Institution is much more competent to any investigation that the Public Service may require ... than the insulated experiment of an Individual Chemist possessing only a private Laboratory.[105]

The proposal was not, however, enough to resolve immediate financial difficulties, and the Managers were finally forced to sell roughly £4,000 worth of 3% and 4% annuities.[106] The Managers' Minutes during this period reveal a preoccupation with various 'Plans of providing for the deficiency of Income' and the desire that the RI 'be converted into a Public and National Establishment divested of any Rights of private Property ...'.[107] As one Committee report put it, 'it is impossible to go on with the least hope of success under the present Charter & Constitution'.[108]

The second step in the proposed revision was more fundamental, and involved the passing of an Act of Parliament which dissolved the hereditary Proprietorship and substituted a single class of membership, thereby putting the control of the Institution into the hands of all who belonged to it.[109] Hereditary Proprietorships, the Managers declared, 'have operated as Impediments to the general Interest and Co-operation which appear to be essential to such a Publick National and Permanent Establishment ...'.[110] They also decided to abolish the rotational system of elections for three-, two-, and one-year Managers and Visitors, and instead ruled that the term of these offices would be for one year only, and that a Manager or Visitor could not be elected for two consecutive terms of office. Finally, the Managers abrogated their own power to elect new Proprietors—now simply called Members—and gave this privilege to the Members alone.

In order to explain the changes to the Proprietors, the Managers requested Davy to deliver a lecture on the subject, which he did on the evening of 3 March 1810. Davy outlined the original plan of the Institution, and explained—though not very completely or convincingly—why the RI had failed in some of its goals. He described the proposed changes, and told his

---

[105] RI Archives, Box 15 (old numbering system), Folder 128, dated 1 May 1809. Other replies preserved in this folder are from the Boards of Admiralty, Agriculture, Customs, Excise, and Ordnance, as well as from the Lord Warden of the Stanneries and the Surveyor General of the Woods and Forests.

[106] *MM*, IV, 474 and 476.

[107] ibid., pp. 461 and 469.

[108] ibid., V, 15.

[109] This was done by Act of Parliament 50 Geo. III, c. 51, 18 April 1810. The Act was steered through the House of Commons by Sir John Sinclair.

[110] From the text of the Bill.

audience that the Proprietors were 'giving up their private interests for the purpose of founding what may be called a National Establishment . . .'.[111] To reassure the original founders of the RI, he quickly added, to 'the great Landed Proprietors of the country, such an Institution cannot fail of being highly useful, even as far as their mere common interests are concerned'. They may rest assured, he continued, that '[w]hatever specimens they may send, will be carefully examined and reported upon'. Nevertheless, he concluded, the restrictive policy is over. 'Our doors are open to all who wish to profit by knowledge . . . .'[112] It was far from being so straightforward as that. Davy's speech also praised the selective policy of the RI, and the vagueness and equivocation of the address reflected the cross-purposes of the Institution which the Managers were struggling to resolve.

Nevertheless, as indicated, the Act of 1810 did embody three important measures. The abolition of hereditary Proprietors was designed to open the doors to a larger clientele. A Proprietor would now be a Member for life, and could, in his will, designate his wife, child, or any blood relation a Life Subscriber; but the automatic connection to the RI ended after that.[113] The second change, the opening up of the election system, was much more significant, and illustrates what sociologists have termed the 'circulation of elites'. As of 1 May 1811, and every first of May thereafter, the Members would elect from amongst themselves fifteen or more Managers and fifteen or more Visitors, the major part of each group not having held either office the previous year.[114] The implications of this are obvious: if more than 50% of each set of annually elected governors had not served the previous year, the control of the Institution by the 'old guard' would necessarily be attenuated. And the new system did have an immediate impact. If we compare the fifty governors of 1799–1811 with (say) the seventy-three governors of 1811–16, we find a sharp drop in representation of the landed interest, from 68% to 46%. This was not accomplished by a drop in absolute numbers, however; roughly the same men who represented the landed interest during 1799–1811 served as governors during 1811–16 as well. Instead, the drop is the result of a broader base. Since the new bylaws did not permit office holding for two consecutive years, greater numbers were needed to fill vacancies. In other words, the immediate impact of the policy change was the influx of new, and different, personnel.

The changes of 1809–11 brought the agricultural phase of the Royal Institution to a close. Not that this occurred overnight. The link with the Board was not severed, as we would expect from an institution still dominated by the landed interest. Davy published the *Agricultural Chemistry* in 1813, and continued to attend Board meetings down to 1817. In 1819 Professor John Millington delivered a course of lectures on agricultural machinery, and in

---

[111] Davy, *Lecture, on the plan which is proposed to adopt for improving the Royal Institution, and rendering it permanent* (London: William Savage, 1810), p. 13.

[112] ibid., pp. 13–14.

[113] Any Proprietor who did not wish to be made a Member was remunerated for the loss of status, which an actuary employed by the RI valued at £42 (*MM*, V, 59–60).

[114] *MM*, V, 22–3.

1820, as we saw, John Cox Hippisley offered the Board a home with the RI when the former was threatened with dissolution. Yet one cannot help thinking that events of this sort were now more symbolic than real. The deliberate use of science for specifically agricultural projects, which had characterized the first decade, was absent during the second; and as early as 1810 the Managers found it necessary to cancel the subscription to *The Agricultural Magazine* because it was no longer being read.[115] The shift in policy was also accelerated by one of England's worst agricultural depressions, which began in January of 1814, and in the face of which the landed interest chose to push through the Corn Law rather than rely on chemistry. Finally, if some of the governors did retain hopes of continuing to use Davy for their private agricultural concerns, they would find that he was no longer strictly an Institution employee. His practical work was turning into national service: ventilating the House of Lords, inventing the miners' safety lamp, and working on the protection of copper sheathing of ships for the Navy, to name Davy's best-known efforts in this direction. The very success he had achieved as a technical problem-solver and as a 'salesman of science' at the RI made him increasingly unavailable for Institutional service. Thus in 1812 Davy ceased lecturing there and continued on as Professor of Chemistry only in an honorary capacity.[116] He was made a Manager and his role at the RI changed from scientist to scientific administrator, although his link with the RI became progressively weaker. Yet the ideology which had motivated the landowners was an enduring legacy, and the emergence of a concrete alternative to the amateur tradition was an event that would not be forgotten. If the landowners had had their day vis-à-vis science, the opportunity was ripe for other classes to begin defining it in accordance with *their* special goals.

[115] ibid., p. 12.
[116] ibid., p. 299.

4
Toward a Rational Society

Let us now recapitulate some of the key developments in the social history of English science which occurred during the transition from an agrarian to an industrial economy, insofar as the Royal Institution was affected by this change. In 1799 the dominant ideology of science was the gentleman amateur tradition. Due to the Baconian revival in the North, this was no longer the only way of regarding scientific activity; but the science club network of the provinces was more an industrial version of the hegemonic tradition than a departure from it. The possession of entrepreneurial attitudes by a small fraction of the aristocracy, however, consolidated the Baconian ideology and made its institutional sustenance possible. At this point the elite did (as Theodore Roszak argued) begin buying up the experts; or, in the case of Davy, creating them. So successful was Davy, both in word and deed, that commercially and industrially backed copies of the RI began forming throughout the country. Science as the key to successful entrepreneurship, and as a commodity in itself, thus became part of the national consciousness.

Within the RI itself, the strain over the control of scientific research led to a serious break, the formation of the London Institution, and very nearly to the collapse of the RI as well. This could never have occurred within the context of the amateur tradition, which was commonly regarded as the 'property' of the aristocracy; but the Baconian ideology was historically derivative from commercial and industrial classes, and was once again seen as salient to their further economic development. There was, then, something to fight about. Furthermore, the RI had constituted itself on an impossibly narrow base. There were simply not enough improving landlords in Great Britain to make its solvency possible. Faced with bankruptcy or attenuation of control, the governors of the RI chose to bend rather than break.

As it turned out, the loosening of electoral control did not bring about greater participation by the commercial, colonial, or manufacturing groups. These groups found it much easier, as the formation of the London Institution shows, to set up their own organizations rather than challenge the RI. Merchants and manufacturers often paid the RI to do laboratory work for them in the years to come, and under Brande's leadership, the Institution did strive to make itself relevant to the concerns of such men; but their attempt to control the gover-.

norship was largely abandoned. By the end of the 1820s, the combined commercial/colonial/industrial control of the Managers and Visitors was roughly 13%.

The group that did come to power in the place of the landed aristocracy was the professional middle class. In fact, between 1799 and 1840 there was a symmetrical role reversal between these two groups within the governorship, with the turning point occurring somewhere around 1824. During the first decade the landed interest held 68% of the governing positions; by the closing years of the 1830s this proportion had dropped to less than 6%. Professionals had a 6% representation during 1799–1811;[1] this changed rapidly after the new election policy in 1811, with a sharp jump to roughly 29%, and then a steady rise to about 60% by the end of the 1830s, with the majority of these drawn from the legal profession.[2]

On the face of it, the desire of the professional classes to control a scientific institution that stood for some combination of entrepreneurship and rational entertainment seems puzzling. Most of the men we are talking about were barristers. Entrepreneurship was foreign to their value system, and barristers were not typically in need of any status enhancement (which governorship of the RI might have been able to provide). Furthermore, if rational entertainment *was* their goal, we would expect to have a large number of lawyers among the Proprietorship (i.e. during the first decade), whereas it amounted to 3.6%.[3] It is thus not immediately obvious what science could do for such men, and their commanding position at the Institution seems rather peculiar.

Part of this dilemma can be resolved by treating the professions as dynamic rather than static groups, i.e. as groups constantly in a state of transition (especially during the early nineteenth century) rather than as immutable sociological categories. Not all lawyers, nor all doctors, conceived of their profession in the same way. The fact that the percentage of professionals among the governors continued to mount steadily, year after year, would strongly suggest that science was becoming increasingly germane to the con-

[1] And even then, the statistics are deceiving, for the three men were Bernard, Hippisley, and Isaac Corry, whose interest at this point was agricultural (and in Hippisley's case colonial as well), but definitely not professional.

[2] This change cannot be explained in simple terms of familial inheritance, i.e. as the result of second-generation governors, younger sons of the artistocracy, entering law and medicine. For one thing, the Act of 1810 provided for an end to hereditary status. The subsequent generation, upon the death of Members, became Life Subscribers, but not Members themselves, i.e. not eligible for election to the governorship. Nor were Members of the next generation (for the most part) offspring of the first group of governors. Second, the dilution of aristocratic control was largely due to the 50% rule, by which no more than half the governors in any one year could hold consecutive terms of office; hence the necessity for completely new personnel. Thirdly, from 1826 the RI experienced a rapid increase in Membership, much of it from the legal profession, from which new governors were subsequently elected.

[3] Legal interest in the Royal Society was also traditionally low, and Arnold Thackray found that very few lawyers were members of the Manchester Literary and Philosophical Society, one of the leading institutions for rational entertainment in the country. *See* his 'Natural Knowledge in Cultural Context: The Manchester Model', *The American Historical Review*, LXXIX (1974), 695.

cerns of some of them. Hence before we can understand the nature of the RI in the post-Davy period, and the impact of the new governorship upon the Institution's scientific activity, we must try to obtain some understanding of the very rapid transformation that affected law, medicine, and government administration during the late eighteenth century and after, and which came to fruition in the link with science that some regarded, by 1820, as essential for these disciplines.

If the Industrial Revolution altered agricultural development in a critical way, its impact on the professions was no less significant, and perhaps more enduring. Prior to about 1750, medicine and law were actually marginal to the nation's economy.[4] The typical doctor was hardly the Lydgate variety depicted in *Middlemarch*—the improving physician, the medical reformer with a strong professional commitment. Rather, he was of the cast which Lydgate eventually found impossible to overcome, and to which, as a doctor at a wealthy spa, he ultimately succumbed. Prior to the late eighteenth century, and well into the first half of the nineteenth century, the doctor served only a small number of wealthy patients. He was first and foremost a gentleman, with technical expertise playing a minor, and knowledge of the classics a major, role in his education.[5] It is even difficult to call medicine a profession before 1800, since physicians were not allowed to charge for their advice but were, in the custom of polite society, 'reimbursed'.[6] The situation could not remain like this for very long. There were demands for medical care from new sectors of society, and with rapid urbanization and overcrowding, problems of health and sanitation were beginning to render the aristocratic model of the polished physician anachronistic. Health reform and the 'sanitary idea' were the opening wedge for the emergence of a very different type of doctor.[7]

Two trends began to develop at this time which went hand in hand. The first, given the new opportunities, was a proliferation of medical services in the form of surgeons and apothecaries.[8] The latter, the lowest branch of the profession, had been little more than shopkeepers or grocers, in whom apothecaries in fact had their historical origins.[9] In the late eighteenth century they were effec-

---

[4] Philip Elliott, *The Sociology of the Professions* (London: Macmillan, 1972), p. 15.

[5] Thus Charles Newman, in *The Evolution of Medical Education in the Nineteenth Century* (London: Oxford University Press, 1957), notes (p. 9) that the examination for Licentiate of the Royal College of Physicians was held in Latin down to 1830. *See also* the testimony by Sir David Barry in the 1834 Committee Report on the Royal College of Physicians (House of Commons, Report from the Select Committee on Medical Education [13 August 1834], Paragraph 2573). Thackray reproduces some typical quotations on the role of the doctor in 'Natural Knowledge in Cultural Context', p. 685.

[6] Newman, *Evolution of Medical Education*, p. 1.

[7] Asa Briggs, '"Middlemarch" and the Doctors', *The Cambridge Journal*, I (1948), 755.

[8] It was estimated in 1795 that for every London patient treated by a physician, twenty were attended by apothecaries (Newman, *Evolution of Medical Education*, p. 81). In the London of 1840, there were roughly fifty physicians, seventy surgeons, and 900 apothecaries and surgeon-apothecaries; *see* E. A. Underwood (ed.), *A History of the Worshipful Society of Apothecaries 1617–1815* (London: Oxford University Press, 1963), p. 192n.

[9] W. S. C. Copeman, *The Worshipful Society of Apothecaries of London 1617–1967* (London: Pergamon Press, 1967), p. 16. Their status was, however, steadily rising during the late

tively functioning as doctors to the poor, as Adam Smith was one of the first to point out.[10] The second trend—and Lydgate typifies this very well— was the use of science as a banner under which the reformers of the new amorphous medical community began to march. As in the case of industry, scientific theory contributed very little to actual practice; but the popular view that the new world of innovation and machinery was a result of the application of science easily spilled over into medicine.[11] For members of the lower branches of the profession seeking to raise their status, now that the Industrial Revolution had created the opportunity, science and improvement were almost synonymous. As Lydgate's story illustrates, the improving physician was not an overnight success, but he was part of an increasingly popular movement that had science, public health, and professionalization tied into one campaign. Medical reformers saw an almost sudden opportunity for a very specialized and prestigious role, and science was, so to speak, the heart of their case. As early as 1804 the apothecaries, who were agitating for the establishment of professional standards, appealed to Sir Joseph Banks to lend his authority to their effort, and succeeded in obtaining it.[12] Little progress was made, however, until 1812, when the apothecaries met in London and formulated a bill to regulate medical practice. As both the Royal College of Physicians and Royal College of Surgeons were opposed, passage was delayed until 1815, when 'the Act for better regulating the Practice of Apothecaries throughout England and Wales' (55 Geo. III, c. 194) became law and established a system of qualification and registration along modern lines. To be licensed by the Apothecaries Society, the student now had to produce certificates of attendance at lectures on anatomy, physiology, chemistry, and materia medica. The Act of 1815 is thus generally credited with raising the level of competence of the profession;[13] but it also served to reinforce the perception that improved status could only be justified in terms of the mastery of scientific theory—even when that theory actually had very little to contribute to medical practice. Faraday, for example,

eighteenth century. *See* Bernice Hamilton, 'The Medical Professions in the Eighteenth Century', *The Economic History Review*, Second Series, IV (1951), 166.

[10] 'Physicians to the poor at all times and of the rich when the danger is not very great' (1790); quoted in Copeman, *Worshipful Society*, p. 53.

[11] Newman, *Evolution of Medical Education*, pp. 56–7 and 99–100. *See also* W. L. Burn, *The Age of Equipoise* (London: George Allen and Unwin, 1964), p. 203.

[12] Newman, *Evolution of Medical Education*, pp. 59–60. The first attempt at regulation came from the Lincolnshire Benevolent Medical Society, of which, as a prominent Lincoln resident, Banks was a patron. He also became a patron of a new group in 1805 known as 'The Associated Faculty'.

[13] ibid., pp. 65–80, and Copeman, *Worshipful Society*, pp. 49–67. For an important dissenting view on the effects of the Act, which sees it as capitulation to the Royal College of Physicians, and therefore a retrograde step, *see* S. W. F. Holloway, 'The Apothecaries' Act, 1815: A Reinterpretation', *Medical History*, X (1966), 107–29 and 221–36. Holloway's view is corroborated by the fact that medical reformer Thomas Wakley attacked the Act, which suggests that it was indeed a compromise rather than real reform. However, Holloway's argument that the Act actually retarded medical education, and was in no way responsible for the establishment of various courses and schools, is less convincing. *See also* his 'Medical Education in England 1830–1858: A Sociological Analysis', *History*, XLIX (1964), 299–324.

wound up giving lectures on electricity to medical students for their certification, to the great puzzlement of Bence Jones, who was then a member of the class (*see* below, p. 135). In short, the emerging technostructure that claimed science as the justification for its own special position—now regarded as a truism—was 'wise before the fact'. Science did not contribute to medicine at this time any more than it contributed to agricultural innovation in the late eighteenth century. It was, rather, an ideological instrument used by medical and agricultural groups, and by extension, became organized as the by-product of these social changes.

In any event, the value of science for medicine became an increasingly popular theme. *The Lancet* argued in 1824 that there was not 'a more important or valuable branch of Medical education than Chemistry . . .. It is . . . now justly regarded as the ground-work of all medical knowledge . . .'.[14] The same year the Apothecaries Society reported that it was

> anxious to give a scientific character to the art, and . . . [a]s a proof of their anxiety to affect this object . . . had appointed Professor Brande to give lectures to pupils at the Hall gratuitously in materia medica, pharmacy, and chemistry . . ..[15]

In general, the improving physician was less interested in polite society than in the world of epidemics and infirmaries, for it was the latter that was moving with the times, and in which he would make his mark. A. B. Granville, a physician who served as a governor of the RI and attacked the Royal Society in *Science Without a Head* (1830), noted that during this time the doctors of London were becoming zealously interested in managing public institutions, hospitals, and scientific academies.[16] Science, a new professional role, and the control of policies relating to sanitation and public health were, by the 1820s, facets of the same issue.

A similar transformation was simultaneously taking place within the legal profession, but this was part of a much larger alteration in pre-Victorian consciousness that comes under the rubric of the 'condition-of-England' question and its response, the so-called 'revolution in government'. Medical professionalization was part of this as well, but much more restricted in scope; for the general implications of widespread crime, disease, and social dislocation meant (to some) that the very nature of English government might have to be reconsidered, and it is through the law that the functions of the state are formally defined. By the 1820s, the social havoc that was part of the rise of a full-

[14] *The Lancet*, 18 March 1824, p. 384.
[15] ibid., 11 January 1824, p. 57.
[16] Paulina B. Granville (ed.), *Autobiography of A. B. Granville* (2 vols.; London: Henry S. King, 1874), II, 50. This difference between image and substance was also remarked upon by the Victorian engineer William Pole (not to be confused with the RI governor of the same name). In his autobiography, Pole stated that in terms of actual practice, science was irrelevant to engineering. He continued: 'When I became engaged on Engineering Work in London I soon found that scientific knowledge was essential for attaining a good standing in the profession . . .'. See William Pole, *Some Short Reminiscences of Events in my Life and Work* (London: privately printed, 1898), pp. 3–4.

blown industrial economy began to reach crisis proportions. 'Britain had become,' writes Kitson Clark, 'a miserable, overcrowded, undergoverned country ....'[17] Agrarian riots continued down to the last Swing uprising (1844), while the cities saw machine breaking and a staggering increase in crimes against property. All of this was popularly emphasized to the point of creating a climate of fear that easily outstripped the reality. The Constabulary Report of 1839 noted that mob action for social and political reasons was becoming common, and documented the details of mob disorder and labour activity 'for the titivation of its timid readers'.[18] The Regency, for all its popular image of spa life and fine society, was one of the most brutal periods in English history. Peterloo (1819) was an indicator of how threatened the custodians of order were, and the Cato Street conspiracy (1820), although tactically foolish, was not an isolated demonstration of resentment of the government's policy of repression. The reaction of the government to the disorders of the period beginning in 1811 included the use of a military force more than six times as large as had ever been needed for such a purpose in the history of England.[19] Strikes, riots, hangings and transportation continued through the 'hungry forties', and Crystal Palace is usually taken as the point at which enough material wealth, or at least the hope of the good life, permeated downward far enough so as to temper the social unrest. Until this time, the ruling class saw itself as being in a near-constant state of siege. Both middle and upper classes viewed themselves as an island in a sea of poverty, eye to eye with a hostile population that somehow had to be kept in check. Even the Managers' Minutes of the RI record that in 1819 the Assistant Porter, dressed in his livery clothes, was attacked and robbed by a mob on its way to a meeting at Covent Garden.[20] For the purpose of public safety, the old machinery of government (the Combination Acts of 1800, for example) seemed, even before the Regency, increasingly inadequate.

It was around the issue of social control that the so-called 'revolution in government', or creation of an administrative and civil service apparatus for enforcing the law and running the state, first began to crystallize. That the law might need a bureaucracy for its enforcement was a strange notion prior to this time. The idea of a preventive policy network had hitherto been regarded as 'Continental' and 'despotic', a dangerous erosion of civil liberties.[21] This did not prevent the government from actually employing a police spy network in

[17] G. Kitson Clark, "Statesmen in Disguise": Reflections on the History of the Neutrality of the Civil Service', in Peter Stansky (ed.), *The Victorian Revolution* (New York: Franklin Watts, 1973), p. 82.
[18] E. C. Midwinter, *Victorian Social Reform* (London: Longmans, 1968), pp. 15–16.
[19] Frank O. Darvall, *Popular Disturbances and Public Order in Regency England* (London: Oxford University Press, 1934), p. 1.
[20] *MM*, VI, 271.
[21] R. J. White, *Waterloo to Peterloo* (Harmondsworth: Penguin Books, 1957), pp. 116, 161, and *passim*; Leon Radzinowicz, *A History of English Criminal Law and Its Administration from 1750* (4 vols.; London: Stevens and Sons, 1948–68), I, 319, and IV, 37; Henry Parris, *Constitutional Bureaucracy* (London: George Allen and Unwin, 1969), p. 164; and Charles Reith, *A New Study of Police History* (London: Oliver and Boyd, 1956), p. 122.

the early nineteenth century, but the sources of information on seditious activities were never openly stated, and the existence of a civilian police network necessarily remained unofficial policy. By 1829, however, when Peel's Metropolitan Police was created, the security of having such a coercive apparatus seemed a far lesser evil than the chaos and anarchy that appeared to lie just around the corner. Much of the legislation of the 1820s reflected the middle- and upper-class belief that an elaborate control apparatus was the only viable alternative. It was the age in which asylum, workhouse and prison came to be regarded as necessary and even reasonable social institutions, rather than extraordinary; as places to put those people who were unable or unwilling to adapt to changes beyond their control. It constituted, in effect, the beginnings of modern industrial society.

It was thus under the threat of social dissolution that the preoccupation with social control arose. The emergence of government by bureaucracy was a reaction to what were perceived to be disintegrating forces. The rationalization (in the Weberian sense) of industrial society is frequently given a progressive or optimistic connotation in terms such as 'improvement' or 'modernization', and it was certainly a process that had more than the control of dissident forces as part of its programme; but by and large improvement and the craving for order conveniently overlapped. The introduction of gas illumination in 1809, for example, was both an attempt to control crime *and* part of the general face-lift which London received during the Regency, and which included street paving, better water supplies, and the architecture of John Nash.[22] The years to come saw an almost absurd obsession with disinfecting and ventilation, as though crime and chaos could be exorcised by enough chlorine and fresh air; and the Royal Institution was often called upon to arrange the requisite scrubbing and airing. These seemingly progressive changes were thus rooted in a morbid anxiety. The emphasis on the external control of the environment, which is a hallmark of modernization, reflected the very real fear that inner control was rapidly slipping away.[23]

[22] On this *see* Fernand Braudel, *Capitalism and Material Life*, trans. Miriam Kochan (London: Weidenfeld and Nicolson, 1973), p. 436, and White, *Waterloo to Peterloo*, p. 23. Nash's role in Regency England is comparable to Haussmann's during the reign of Napoleon III.

[23] Although I occasionally use the term 'modernization' in this discussion, I mean something roughly equivalent to Weber's concept of rationalization rather than the technical sociological treatment provided by such writers as Samuel P. Huntington. Weber characterized the transition to modernity by the replacement of the traditional elements in society by legal-rational ones, and the attraction of bureaucracy as a form of government best able to reduce unpredictability (Durkheim and Tönnies provided similar descriptions). Although this is rather vague, my own reading of the literature suggests the possibility that more has been lost than gained by precision. Huntington's description, for example, could be a convenient rubric under which to discuss changes such as gas lighting and street paving, but I am dubious as to whether anything is actually explained by it (*see* his *Political Order in Changing Societies* [New Haven: Yale University Press, 1968], Ch. 1). The term is used to describe the profound difference between traditional and modern society without giving any interpretation of the causes of change. It is erroneously used interchangeably with 'industrialization' (on this *see* E. A. Wrigley, 'The Process of Modernization and the Industrial Revolution in England', *Journal of Interdisciplinary History*, III [1972], 225–60); and it is used whiggishly, implicitly assuming that the process is universal and

How could such external control be assured? It was this question that brought science and the law together during this time to create what can most accurately be described as a 'legal ideology of science', an ideology that received its finest expression in the writings of England's greatest thinker in the field of jurisprudence and criminal law, Jeremy Bentham. Although the nature of Benthamism, the exact identity of his followers, and their role in policy-making, have been the subject of much debate in recent years, all are agreed that efficiency and expertise were the two major planks in the Benthamite platform.[24] The goal of the Philosophical Radicals, as Phyllis Deane has pointed out, 'turned out to be not freedom from government but freedom from inefficient government ...'.[25] Bentham's real enemy was arbitrariness and amateurism. Just as disease and overcrowding created a new role for the doctor, that of health officer or medical administrator, the presence of chaos and violence created an equally new role for barristers, legislators, and other government or legal personnel: that of social engineer. As Harold Perkin has noted, Bentham was 'the apotheosis of the professional ideal', for it was the new professional class which stood to benefit from his vision of an ad-

therefore teleological, with Western Europe or the U.S. being the climax of development (cf. the discussion in the S.N. Eisenstadt, 'Studies of Modernization and Sociological Theory', *History and Theory*, XIII [1974], 225–52). It is, in short, a loaded term, evoking a positive image of progress and prosperity based on a timeless type of yardstick. Much of Marx's most telling criticisms of classical political economy, notably its ahistoricity, can be levelled against the theory of modernization as well.

The literature is so vast that perhaps a reference to bibliographies on the subject would be more useful: John Brode, *The Process of Modernization: An Annotated Bibliography on the Sociocultural Aspects of Development* (Cambridge: Harvard University Press, 1970), and Allan A. Spitz, *Developmental Change: An Annotated Bibliography* (Lexington: University of Kentucky Press, 1969). The footnotes in Eisenstadt are also helpful.

[24] For general summaries of the debates and relevant bibliography the following sources are useful: A. J. P. Taylor, *Laissez-faire and State Intervention in Nineteenth-century Britain* (London: Macmillan, 1972); Valerie Cromwell, 'Interpretations of Nineteenth-Century Administration: An Analysis', *Victorian Studies*, IX (1966), 245–55; and Roy MacLeod, 'Statesmen Undisguised', *The American Historical Review*, LXXVIII (1973), 1386–1405. Relevant works in the debate itself include A. V. Dicey, *Lectures on the Relation between Law and Public Opinion in England during the Nineteenth Century* (London: Macmillan, 1905); Elie Halévy, *The Growth of Philosophic Radicalism*, trans. Mary Morris (London: Faber and Faber, 1972 reprint); J. B. Brebner, '*Laissez-faire* and State Intervention in Nineteenth-century Britain', in E. M. Carus-Wilson (ed.), *Essays in Economic History*, III (1962), 252–62; Jacob Viner, 'Bentham and J. S. Mill: the Utilitarian Background', *The American Economic Review*, XXXIX (1949), 360–82, and 'Intellectual History of Laissez Faire', *The Journal of Law and Economics*, III (1960), 45–69; W. H. Coates, 'Benthamism, Laissez Faire, and Collectivism', *The Journal of the History of Ideas*, XI (1950), 357–63; David Roberts, 'Jeremy Bentham and the Victorian Administrative State', *Victorian Studies*, II (1959), 193–210; Oliver MacDonagh, 'The Nineteenth-Century Revolution in Government: a Reappraisal', *The Historical Journal*, I (1958), 52–67; Henry Parris, 'The Nineteenth-Century Revolution in Government: A Reappraisal Reappraised', *The Historical Journal*, III (1960), 17–37; L. J. Hume, 'Jeremy Bentham and the Nineteenth-Century Revolution in Government', *The Historical Journal*, X (1967), 361–75; and Jenifer Hart, 'Nineteenth-Century Social Reform: A Tory Interpretation of History', *Past and Present*, No. 31 (July 1965), 39–61. Recent studies include W. H. Coats (ed.), *The Classical Economists and Economic Policy* (London: Methuen, 1971), and Gillian Sutherland (ed.), *Studies in the Growth of Nineteenth-Century Government* (London: Routledge and Kegan Paul, 1972).

[25] Phyllis Deane, *The First Industrial Revolution* (Cambridge University Press, 1965), p. 215. *See also* White, *Waterloo to Peterloo*, p. 136.

ministrative apparatus to perform specialized functions relating to public welfare.[26] The 1820s thus witnessed the appearance of the 'rootless expert', or what Dickens bitterly termed the 'commissioners of fact'; a group whose allegiance was to a professional community rather than to village or family, and who sought to replace the social functions of the latter with the specialized structures administered by that new community.[27]

It is this category (which includes the improving physician) that I have labelled 'Utilitarian' in trying to understand the RI governorship during the post-Davy period; but we shall not, at the RI or in England in general, be able to pinpoint this group in a precise sociological way. Whig reformers, for example, have to be included, despite the fact that the Whigs and the very tiny group (about twenty) of self-proclaimed Philosophical Radicals were often at loggerheads. Agitation over issues such as the Reform Bill tended to conceal their differences, and Benthamites occasionally won elections with Whig support. With the fall of the Whig government in 1834, Whigs and Radicals often found themselves together on the side of reform; the latter appeared, in the public eye, to be a branch of the former after this date.[28] Thus Benthamite ideology was something of a blurry spectrum, and should not be thought of as 'a highly developed body of rigid doctrine held in common by a tightly-knit and severely disciplined sect'.[29] 'The Utilitarians,' as Leslie Stephen put it, 'had no great talent for cohesion.'[30] The best that can be said is that the movement's followers were a group of men with a certain disposition toward problems, which they regarded, in varying degrees, as tractable via quantification and the statistical accumulation of facts. Thus Kitson Clark writes that

> at that moment there were in Britain an unusually large number of people whose habits of mind fitted them for this task, men who could be trusted to apply to any problem careful systematic observation and self-confident and rigorous argument, working from relatively simple and superficial first principles and not confused by too profound or intrusive a philosophy. The followers of Jeremy Bentham were such men, but to call all such men Benthamites is certainly to narrow too much to one group a much more generalized contemporary cast of mind.[31]

This cast of mind can certainly be called 'scientific', but not in the aristocratic or entrepreneurial sense of the term. We are dealing, rather, with a very

[26] Harold Perkin, *The Origins of Modern English Society 1780–1880* (London: Routledge and Kegan Paul, 1969), p. 269. The self-serving nature of the demand for expertise is discussed on pp. 252–88.

[27] Cf. the discussion of Wiebe in the Introduction and R. K. Webb, *Modern England* (New York: Dodd, Mead, 1968), pp. 241–2.

[28] Joseph Hamburger, *Intellectuals in Politics* (New Haven: Yale University Press, 1965), pp. 114 and 153.

[29] A. J. P. Taylor, *Laissez-faire and State Intervention*, p. 33. J. S. Mill, for example, expressed great admiration for men like Coleridge and Carlyle.

[30] Leslie Stephen, *The English Utilitarians* (3 vols.; London: Duckworth, 1900), II, 31.

[31] Clark, 'Statesmen in Disguise', p. 83. Unfortunately for my subsequent analysis, this is as tight as the definition of this category can be drawn. It would be convenient to have as coherent a group as the Board of Agriculture to work with, but professional reformers were much more diverse than agricultural improvers.

different notion of science, one which was absolutely necessary if the new role of professional expert was to be justified. This legal ideology of science was the belief that science was a tool by which to construct an ordered society; that society was not an organism but a machine; that its stability could be arranged or managed through technical know-how; and therefore that the chaos was apparent rather than fundamental.

The best example of this ideology was the Statistical Society, an organization dominated by the Utilitarians, to which a number of the Royal Institution's governors belonged. This included its first President, the Marquis of Lansdowne, who was active in the RI governorship from 1811 to 1836. Founded as an offshoot of the British Association for the Advancement of Science in 1834, the Statistical Society was committed to the amelioration of social evils via the compilation of factual information. Its goal, as Philip Abrams has remarked, was 'to end politics; in the sense of obliterating conflicts of principle'.[32] Appoint enough Select Committees and Royal Commissions, compile enough data, and the answers will fall out of the material as if by sheer weight.[33] Thus the Society's original constitution contained the following paragraph:

> The Statistical Society will consider it to be the first and most essential rule of its conduct to exclude carefully all Opinions from its transactions and publications—to confine its attention rigorously to facts—and, as far as it may be found possible, to facts which can be stated numerically and arranged in tables.[34]

England, on this view, did not have social problems. Rather, it had 'technical difficulties', and thus the various facets of the condition-of-England question and the modernization of the nation were scientific matters, amenable to scientific treatment. Thus was born, as a political force, the technological fix. This value judgment of the Society was so taken for granted by its members that it excluded any possibility of their recognizing its ideological nature; and in particular, it ruled out any recognition that the 'technical difficulties' to be solved by data collection—crime, poverty, disease, pollution, and the like—were endemic to the structure of industrial society itself rather than 'unnecessary' by-products of it. Although hardly a 'conspiracy' on the Society's part, one suspects that this view was not accidental. One of the central functions of an ideology—whether in terms of the strain theory or the interest theory (*see* the Introduction, p. *xxiii*)—is to mask the actual motives that underly it, and which it ultimately serves to fulfill. The failure to recognize the endogenous character of industrial problems not only gave the social engineers their *raison d'être*; it also created solutions which, unsurprisingly, did not solve very much. When Carlyle wrote that the Poor Law was designed to cure pauperism rather than poverty, he unwittingly stated what was a metaphor for all scientific treatment of

---

[32] Philip Abrams, *The Origins of British Sociology: 1834–1914* (Chicago: University Press, 1968), p. 13.

[33] ibid., pp. 5–9. *See also* J. F. C. Harrison, *The Early Victorians* (London: Cox and Wyman, 1971), Ch. 3.

[34] Abrams, *British Sociology*, p. 15.

social problems.[35] These 'solutions' were in fact anodynes, and their epistemology circular. That is, the Utilitarian definition of science was a result of professional motivations which were supposedly justified by new social problems that were in turn amenable to 'scientific' solutions devised and administered by the professional classes. The real function of this sort of science was to contain problems, not erase them. The mountains of data generated by this ideology thus constituted 'massive but intellectually sterile levers of social reform'.[36] But, as in the case of medicine, it was the image of science rather than its efficacy that was the issue.[37]

Two other organizations that had a strong Benthamite colouration were the London University and the Society for the Diffusion of Useful Knowledge, both founded in 1826. Their overlap with the Royal Institution was substantial. It was no accident that Faraday was offered a professorship at the University as early as 1827, that Brande served as an examiner there, or that the Diffusion Society chose a number of its authors from amongst the RI governorship. The organizations were dominated by the outlook of men like Henry Brougham, and their aims included both the creation of social leverage for the new professional class, and the containment of social unrest. University College, as it was later called (1836), was a deliberate attempt by the Whigs and Utilitarians to break the grip of Oxbridge on education and the training of an elite, and to create an elite of their own.[38] As might be expected from our previous discussion, the University purposely chose to emphasize the sciences. There was a preponderance of chairs in science and medicine. Bentham also intended that the college devote much of its attention to social science and technology, and 'both [he] and the founders of this College were acutely aware

[35] Harrison, *Early Victorians*, pp. 82 and 84.

[36] Abrams, *British Sociology*, p. 5.

[37] It will be understood that I am not arguing that the statistical method, and the anodyne psychology which accompanied it, were the sudden inventions of the nineteenth century. The seventeenth century witnessed a great interest in such issues, as the names of Gregory King, William Petty, John Graunt, *et al.* would indicate. With Petty, as with Bentham, quality control ('political arithmetic') is raised to the status of a philosophy, although for Bentham this mode of analysis becomes identical with truth itself. More to the point is the difference in timing. The early attempt to deal with social problems in terms of measurement and functional analysis never got off the ground. As George Rosen has pointed out, these ideas were not able to take root until the Industrial Revolution, and in the case of public health, not until the 1820s and 1830s ('Economic and Social Policy in the Development of Public Health', *Journal of the History of Medicine and Allied Sciences*, VIII [1953], 406–30, and 'Problems in the Application of Statistical Analysis to Questions of Health: 1700–1880', *Bulletin of the History of Medicine*, XXIX [1955], 27–45). English administration remained strongly parochial into the early nineteenth century, since in Petty's day, and throughout the eighteenth century, there was no class that could benefit from a scheme of national reification. By the time of Bentham's death, however, such a class had appeared, and the simultaneity of statistical organizations is no surprise. Lord Auckland established a Statistical Office at the Board of Trade in 1832; the Manchester Statistical Society was founded in 1833.

[38] H. Hale Bellot, *University College London 1826–1926* (London University Press, 1929), and *Statement by the Council of the University of London, explanatory of the Nature and Objects of the Institution* (London: Richard Taylor, 1827).

of the claims of scientific and technical education . . .'.[39] As at Oxbridge, Greek, Latin, and mathematics were part of the College curriculum, but its founders also provided for courses in political economy, chemistry pure and applied, medicine (including clinical work), engineering and jurisprudence. The prospectus specifically argued for the importance of professional education.[40] 'Particular stress,' writes Hale Bellot, 'was laid upon legal and medical education,' and the founders stated that they wished to attract 'the gentlemen who hold places in the offices of Government'.[41] The medical school was to be linked to a hospital and dispensary, and courses framed so as to meet the requirements of the Royal College of Surgeons and the Apothecaries Society. The promoters of the University quite candidly regarded the medical school as part of the general push for the reform of medical education.[42]

The similarity between these goals and those of the RI during these years (as will be discussed) is not surprising, given the degree of overlap. The first Council of the University included five RI governors, and its shareholders fourteen.[43] Leonard Horner, who had been a governor in previous years, was made Warden and Secretary. John Millington was appointed Professor of Engineering and the Application of Mechanical Philosophy to the Arts (he resigned, however, in 1828). William Ritchie, who also taught at the RI, held the chair of Natural Philosophy and Astronomy from 1831 to 1837, the year of his death. Alfred Ainger was the architect of the University hospital, and Isaac Lyon Goldsmid and the Duke of Somerset were two of the University's heaviest financial backers, and very active in its affairs (it was Goldsmid who acquired the Gower Street site). S. J. Loyd served as auditor of the University's accounts.[44]

Turning to the Diffusion Society, it fitted into a second aspect of the Utilitarian programme, which was the mitigation of working-class dissent via the diffusion of cheap reading material: the *Penny Magazine*, the *Penny Cyclopaedia*, the *Journal of Education*, the *Library of Useful Knowledge*, the *Library of Entertaining Knowledge*, the *Working Man's Companion*, the *Biographical Dictionary*, etc. As the only full-length study of the Diffusion Society reveals, the goal was to bring harmony to the lives of those adversely affected by the Industrial Revolution; and as in the case of the Statistical Socie-

[39] J. H. Burns, *Jeremy Bentham and University College* (London: The Athlone Press, 1962), pp. 8 and 10.

[40] *Statement by the Council*, pp. 10–11 and 31–41.

[41] Bellot, *University College*, p. 54.

[42] ibid., pp. 54, 143, and 145.

[43] The Council members were Viscount Dudley and Ward, I. L. Goldsmid, the Marquis of Lansdowne, Lord John Russell, and Henry Warburton. The additional shareholders were James Alexander, the Duke of Bedford, Samuel Boddington, John Bostock (later a Council member), Nicholas Fazakerley, Lord Gower, Henry Hallam, Charles Shaw Lefevre, and S. J. Loyd. *See Statement by the Council.*

[44] Bellot, *University College*, p. 152; *Statement by the Council*, pp. 17 and 21; University College, London, College Correspondence, letters from Somerset, and Item 2580, letter from Loyd to Thomas Coates.

ty, religion and politics were to be eschewed, and 'useful knowledge' dis-
seminated in their stead. The ultimate function of the 'Sixpenny Science Company', as it was called, was 'to rivet rather than release the shackles on the industrial worker'.[45] In the context of the 1820s it was a significant break with
Tories, Church, and aristocracy, who regarded a little knowledge as a
dangerous thing, even if, in the long run, the differences between Tories,
Whigs, and Utilitarians as regards the lower classes do not seem to have been
very great. Debates were, of course, acrimonious, but Brougham stated quite
flatly in his plans for the Society that the effect of such education would be a
stabilizing one. Natural philosophy, it should be pointed out, was to be given
foremost place in the plan for the *Library of Useful Knowledge*.[46]

As in the case of the London University, there was a high degree of overlap
with the Royal Institution, but then the University and the Diffusion Society
were virtually the same organization, as *The Athenaeum* pointed out in 1833.[47]
The Committee of the Society during its first decade included the following RI
governors: J. F. Daniell, I. L. Goldsmid (who served as third and last
Treasurer), Henry Hallam, Leonard Horner, J. W. Lubbock, R. I. Murchison,
P. M. Roget, Lord John Russell, and Henry Warburton. Eleven other governors
were Annual or Life Subscribers of the Society.[48] Goldsmid, J. F. Daniell,
Roget, Murchison, and Russell were on the Committee for the *Penny
Magazine*, and the first three of these also worked on the *Penny Cyclopaedia*.
Lubbock and Goldsmid helped edit the *Journal of Education*, and Goldsmid,
Murchison, Roget and G. T. Staunton were on the Committee of the
Biographical Dictionary. RI governors who were authors for the Society press
include Millington, who did a volume on hydraulics, and who revised Alfred
Ainger's manuscript on the steam engine (Brunel was also slated for a work on
this subject); Roget, who wrote several volumes on electricity, galvanism,
magnetism, and electromagnetism; J. F. Daniell, who did the volume on
chemistry; and George Long, who was the co-author of a book on geography.
Murchison was signed up for a volume on the statistics of geology, mines, and
minerals. The Society was also eager to enlist Faraday as an author. 'I have over
& over again persuaded Faraday to write for us,' J. F. Daniell told Thomas

---

[45] Monica C. Grobel, *The Society for the Diffusion of Useful Knowledge 1826–1846* (Ph.D.
Dissertation, University of London, 1932). I am grateful to the authoress for permission to quote
from this thesis. How successful the Diffusion Society was in this cooptation is a moot point; it
certainly did not accomplish it single-handedly. *See* Webb, *Modern England*, p. 182.

[46] Grobel, *Society*, pp. 9 and 156. Brougham's ideas were put forth in his *Practical Observations upon the Education of the People* (1825).

[47] *The Athenaeum*, 23 February 1833, pp. 121–2. The information that follows is taken from
the *Penny Magazine*, VII (1838); *Prospectus, the Society for the Diffusion of Useful Knowledge*
(London: William Clowes, n.d.); *Society for the Diffusion of Useful Knowledge* (London: George
Taylor, 1832); *The Biographical Dictionary of the Society for the Diffusion of Useful Knowledge*
(4 vols.; London: Longman, Brown, Green, and Longman, 1842); Grobel, *Society*; and the Diffusion Society Letters at University College, London.

[48] The additional eleven were Alfred Ainger, Samuel Boddington, the Duke of Bedford, E. R.
Daniell, Nicholas Fazakerley, Sir Henry Halford, Charles Holford, S. J. Loyd, Edward Magrath,
Sir John Sebright, and N. A. Vigors.

Coates, the Society Secretary, 'but his time is so fully occupied that he cannot.'[49]

There was, then, in the 1820s, a general rise in professional consciousness that affected medicine, education, and government administration, had a Benthamite view of science as part of its ideology, and found its outlet in organizations such as the Statistical Society, the London University, and the Diffusion Society. To conclude our general survey of this phenomenon, it should be added that the law was similarly affected, and that a number of barristers beyond just Brougham and Romilly were attracted to Utilitarianism and social reform, despite the typically leisured existence of the senior branch of the legal profession.[50] This typicality needs to be stressed. Law, like medicine, enjoyed an essentially relaxed role in British society during most of the eighteenth century. The tasks performed were largely related to the land and the needs of landowners—problems of entail, strict settlement, and inheritance—and required familiarity rather than training or technical expertise. By 1800, barristers did no conveyancing at all. The Bar was essentially a respectable way of life for wealthy gentlemen, frequently younger sons of the gentry or aristocracy, and the Inns of Court little more than residential clubs.[51] The gap between barristers and solicitors, which was very large to begin with, was widened with the coming of industrial society, for new opportunities were created for solicitors—problems of mortgages, stocks and bonds, patents, new companies, transport construction, etc.—that tended to render barristers increasingly irrelevant.[52] Whereas solicitors formed the Law Society in 1825, and won the right to qualify members by examination in 1836, the Bar had no need

[49] University College, London, Diffusion Society Letters, Daniell to Coates, 11 February 1831.

[50] The actual impact of the Industrial Revolution and its associated changes on the legal profession is difficult to chart, primarily because of the absence of any studies of barristers or solicitors during the nineteenth century. This is a curious gap, as Brian Abel-Smith and Robert Stevens point out in their preface to *Lawyers and the Courts* (London: Heinemann Educational, 1967), especially since lawyers became a very influential group after 1820. There is nothing comparable to Wilfred R. Prest's *Inns of Court under Elizabeth I and the Early Stuarts, 1590–1640* (Totowa, N.J.: Rowman and Littlefield, 1972), or Robert Robson's *Attorney in Eighteenth-Century England* (Cambridge University Press, 1959). At this stage most of the information available on legal practitioners in Victorian times is from literary sources such as *Bleak House* or *Pickwick Papers* which capture an important public attitude but are by their nature highly impressionistic (on these *see* William S. Holdsworth, *Charles Dickens as a Legal Historian* [New Haven: Yale University Press, 1928]). For lack of any monographic studies on Victorian barristers, the sketch below is necessarily brief and tentative.

[51] Abel-Smith and Stevens, *Lawyers and the Courts*, pp. 25 and 64 ff.; 'The English Bar and the Inns of Court', *Quarterly Review*, CXXXVIII (1875), 139; *Law Review*, XVI (1852), 183; Thomas Ruggles, *The Barrister: or, Strictures on the Education Proper for the Bar* (2nd ed.; London: W. Clarke, 1818), esp. pp. 23–5; and Frederick von Raumer, *England in 1835* (3 vols.; London: John Murray, 1836), III, 259.

[52] Abel-Smith and Stevens, *Lawyers and the Courts*, p. 19. Michael Birks, *Gentlemen of the Law* (London; Stevens and Sons, 1960), p. 3, states that the solicitor of ca. 1800 was viewed in the same category as a tradesman. A revealing comparison of the gap can be obtained by comparing Samuel Warren, *A Popular and Practical Introduction to Law Studies* (London: A. Maxwell, 1835), which is for barristers and emphasizes a knowledge of the classics, with [Sir George Stephen], *Adventures of an Attorney in Search of a Practice* (London: Saunders and Otley, 1839), a sort of plumber's handbook for solicitors.

of status enhancement nor any interest in educational reform.[53]

Given the small number of barristers at this time—821 in 1814[54]—the high percentage of barristers among the Royal Institution's governors is itself something of a phenomenon. For some, science may have been nothing more than rational entertainment. Yet Bentham's impact upon jurisprudence, the affinity between the law and the administrative 'revolution' previously discussed, and the careers of men like Brougham and Romilly would indicate that the Bar had been at least partially penetrated by Whig reformism and Utilitarianism. Law was, in fact, becoming the chief vehicle of social engineering.[55] Barristers were certainly involved in the social legislation generated by the system of Select Committee and Commission of Inquiry; and in the case of the latter, if it were transformed into a permanent administrative commission, barristers were usually appointed to it.[56] Sydney Smith, a journalist and Whig reformer himself, found this correlation annoying, calling the barrister of six years' standing the government's 'favourite human animal'. 'The whole earth,' he continued, 'is, in fact, in commission, and the human race, saved from the flood, are delivered over to the barristers of six years' standing.' Commission and barristration, he concluded, have become 'the real secret of life . . .'.[57] And it was through these rudimentary forms of bureaucracy that almost every phase of English social life was scrutinized during the first half of the nineteenth century.[58]

We began this Chapter with a discussion of the statistics of the professional takeover of the RI governorship. The foregoing discussion of the changing nature of the professions during this time, however, would suggest that the category of 'professional' is sociologically meaningless. From the mere fact that an individual can be classified as a lawyer or physician it does not follow that he is also a Utilitarian. The real issue is what type of barrister or medical man he actually was. In a number of cases, the fact that a governor was a lawyer or doctor is the only piece of data available; in others, it is clear that social

---

[53] Abel-Smith and Stevens, *Lawyers and the Courts*, pp. 53–4 and 64–8.

[54] W. J. Reader, *Professional Men* (New York: Basic Books, 1966), p. 46. Compare the data given by H. Byerley Thomson in *The Choice of a Profession* (London: Chapman and Hall, 1857), p. 96:

1810— 880	1840—1,835
1821— 820	1850—3,268
1830—1,129	1855—4,035

These figures include men who were no longer resident or practising in England.

[55] Vilhelm Aubert (ed.), *Sociology of Law* (Harmondsworth: Penguin Books, 1969), p. 10.

[56] H. D. Clokie and J. W. Robinson, *Royal Commissions of Inquiry* (Stanford: University Press, 1937), pp. 83 and 93. Select Committees were distinguished from Royal Commissions of Inquiry in that the former were composed of Parliamentarians only, who would on occasion interview various experts. The Commissions, on the other hand, were often composed of the experts themselves.

[57] ibid., p. 93. This was part of an attack on the Ecclesiastical Commission.

[58] Between 1800 and 1833, over 500 Committees presented reports to Parliament. Sixty Royal Commissions of Inquiry were held between 1800 and 1831, and an additional ninety-one between 1831 and 1844. *See* Clokie and Robinson, *Royal Commissions*, pp. 58–9, 62, and 73n.

engineering played a major part in his interests. We shall advance, then, as a criterion for classification in the Utilitarian group (in addition to those few men who were obviously cast in the role of scientific administrators) any one of the following conditions:

(a) Work as an improving physician, i.e. infirmary work, medical administration, public health, and the like.

(b) Membership in the Statistical Society or Political Economy Club, two groups known for their Utilitarian association.[59] This is not foolproof: e.g., Sir Henry Halford, President of the Royal College of Physicians, who belonged to the Society, can hardly be regarded as a professional innovator; but his case was very exceptional.

(c) Active participation in known Utilitarian projects, such as the London University or the Society for the Diffusion of Useful Knowledge.

(d) Work on commissions or other projects obviously connected with the administrative improvement of social conditions, or with administrative reform.

On this basis, the following table can be constructed:

Time Period	Number of Governors	Number of Utilitarians	Percentage of Utilitarians
1811–16	73	9	12.3%
1816–21	77	13	16.9%
1821–26	80	19	23.3%
1826–31	75	28	37.3%
1831–36	72	27	37.5%
1836–41	72	31	43.0%

The Utilitarian component of the governorship rose steadily in the three decades after 1811, with a sharp increase shortly after 1826. By the third quinquennium, its influence was already significant; during 1826–31 it became dominant. These were men who were unusual in the pursuit of professional goals, active on commissions, in issues of public health and penal reform, and so on. However, these figures represent the minimum number, and there are reasons to believe that the actual figures are much higher. If Benthamism was a cast of mind, we would guess that there were a number of Utilitarians who were not active as such, but no less ideological in their viewpoint. There were also nineteen barristers who could not be further identified as Utilitarians, but whose work (in a few cases) may suggest it.[60] Richard Barnewall was a court reporter, and Edward Sterling a *Times* journalist; William Sturges Bourne and Benjamin Hall were both very ac-

[59] *See* S. E. Finer's essay in Gillian Sutherland, *Studies in the Growth of Nineteenth Century Government*, pp. 15–19.

[60] Barnewall, Berens, Bourne, Burroughs, Erskine, Hall, Harrison, Hovenden, Jordan, Leigh, Moore, J. Nicholl, Rudge, J. Russell, Rust, Sterling, Templeman, Thompson, and Warry.

tive politically, and Lord Erskine, who once took the trouble to ridicule the Royal Society in 1810 in Parliament for its pursuit of irrelevant knowledge, joined with Romilly in an attack on the corruption of the Court of Chancery. It is always possible, as we have indicated, that barristers joined the RI for rational entertainment; but one could find it without the RI, and certainly without the cost of Membership. The Utilitarians were struggling for social legitimation in the 1820s and 1830s; wealthy barristers of the old school were not. Finally, the RI was the first scientific institution of any kind that had a large number of barristers running it, which suggests that something more than a gentleman's interest in the subject was at work.

As with the agricultural interest during the first decade, it will be instructive to obtain a qualitative impression of the men who shaped the direction of the RI in the later period. A good example of the improving physician is John Bostock, who wrote his thesis at Edinburgh on medical chemistry, then went on to publish on subjects such as galvanism and pharmacology, contributing to Brewster's *Encyclopaedia* and the *Philosophical Transactions*. In Liverpool he worked at a public dispensary, and helped to set up fever wards. Bostock also co-founded the Royal Institution of Liverpool in 1814, acted as first professor, set up its laboratory and joined it to the medical school of the Liverpool Infirmary. When he moved to London in 1817 he began lecturing on chemistry at Guy's Hospital and became active in a number of scientific societies. In the late 1820s he was asked to serve on the Metropolitan Water Commission and had to deal with the problem of purifying the Thames. At this time also he became a member of the Council of the London University and very active in University affairs. Bostock died of cholera in 1846.[61]

Augustus Bozzi Granville also acted in the capacity of a chemistry lecturer, succeeding John Davy and preceding Brande at the Windmill Street medical school. He introduced the use of prussic acid for chest infections and established an infirmary for sick children. Today he is probably best known to historians for his attack on the Royal Society, which focused on the non-scientific nature of its membership, but during his lifetime he was famous as a vigorous medical reformer, in which capacity he displayed a typically Benthamite approach. He attacked contemporary standards in midwifery and the lack of professionalism in obstetrics; spoke out against English laws as inadequate to distinguish between professional physicians and quacks; and concerned himself with issues of water purity and epidemiology. His *Report on the Practice of Midwifery at the Westminster General Dispensary* (1819) characteristically tabulated data on pregnancies, disease, prescriptions, mortality, etc., and his book *Sudden Death* (1854), which dealt with death by apoplexy and paralysis, was based on data in the office of the Registrar-General. He also compiled obstetrical statistics on the working class, and testified before Parliament on the improvement of the Thames. Granville was

[61] S. G. Smith, *The Contributions to Science of John Bostock, M.D., F.R.S., 1773–1846* (M.Sc. Dissertation, University of London, 1953).

13a. John Bostock

13b. Peter Mark Roget

13c. Henry Warburton

13d. I. K. Brunel

Secretary of the RI Visitors between 1832 and 1852 and claimed responsibility for keeping the Institution out of debt.[62]

Peter Mark Roget, who served as Fullerian Professor of Physiology at the RI (1834–7), was possibly the most versatile man of the age. Besides publishing the *Thesaurus* (itself a model of codification) and inventing a slide rule and the pocket chess set, Roget was a prolific scientific writer and lecturer. As the nephew of Romilly, he came to Utilitarianism quite naturally. He spent six weeks with Bentham at one point, after the latter consulted him on a scheme for using the sewage of London, and in 1827 he and Brande were commissioned by the government to inquire into the metropolitan water supply, for which task John Bostock did the water analyses. He was physician to the Manchester Infirmary in 1805, and later to the Northern Dispensary in London. He helped found the Manchester Medical School and the University of London (where he served as an examiner), and acted as medical officer for Milbank Penitentiary. As we have seen, he was very active in the Diffusion Society, which he helped found.[63]

The barristers at the RI tend to be much less well known, often disappearing into the grey recesses of administrative work. Edmund R. Daniell, an equity draftsman and conveyancer, served as a court reporter and later (1842–54) Joint Commissioner of the Birmingham Court of Bankruptcy. His handbook on the operations of the Court of Chancery, commonly referred to as 'Daniell's Chancery Practice', ran to eight editions.[64] James W. Farrer served as Master in Chancery between 1824–52, and became interested in the reform of that office, which, he claimed, should ideally run like machinery. His recommendations for greater efficiency were included in a report he submitted to the Equity Committee of the Law Amendment Society. He also collected data on the condition of the working classes in Lancaster, was friends with Edwin Chadwick, and corresponded with him on subjects such as legal reform.[65] Edmund S. Halswell, equity draftsman and conveyancer, served as Metropolitan Commissioner in Lunacy (1836–7) and was an early member of the Statistical Society.[66] Finally, barrister and legal historian Henry Hallam proved to be very interested in the statistics of social distress, and as a founder and the first Treasurer of the Statistical Society he privately financed house-to-house surveys for the Society in the 1830s.[67] He was also on the Committee of the Society for the Diffusion of Useful Knowledge.

[62] Granville, *Autobiography*, II, 214ff.; G. T. Bettany, 'Augustus Bozzi Granville', *DNB*, XXII (1890), 412–14; and Charles F. Mullett, 'Augustus Bozzi Granville: A Medical Knight-Errant', *Journal of the History of Medicine and Allied Sciences*, V (1950), 251–68.

[63] W. W. Webb, 'Peter Mark Roget', *DNB*, XLIX (1897), 149–51.

[64] *A Treatise on the Practice of the High Court of Chancery. See also* the 1830 *Law List*, and the sketch of Daniell in Frederic Boase's *Modern English Biography*, I (1892), 811.

[65] University College, London, Chadwick, Papers; Samuel Miller, *Observations on the necessity for continuous proceedings in The Offices of the Masters in Chancery* (London: S. Sweet, 1848); and the sketch of Farrer in Boase, *Modern English Biography*, I (1892), 1026.

[66] J. A. Venn (ed.), *Alumni Cantabrigienses*, III (1947), 209.

[67] Abrams, *British Sociology*, p. 33.

Utilitarians at the RI also included a number of famous organizers and administrators. One such was the factory inspector, Leonard Horner. Horner was first interested in geology, but soon turned to scientific education, founding the School of Arts in Edinburgh for the instruction of mechanics. Later he served as Warden of London University. As chief factory inspector in Lancashire (1833–60), Horner worked on child labour laws, and provided exhaustive reports on working conditions.[68] Another of the great 'modernizers' was I. K. Brunel, one of England's foremost engineers, who helped build the first Thames Tunnel and then served as chief engineer of the Great Western Railway. He was also a builder of steamships and docks, and did experiments for the Admiralty. Brunel joined the Statistical Society soon after its formation, was active in the Great Exhibition, and intended to write on the steam engine for the *Penny Cyclopaedia*.[69] Sir Isaac Lyon Goldsmid, also an early Statistical Society member, was bullion broker to the Bank of England and closely associated with the Philosophical Radicals and social reform. As already noted, he was the central figure in the first Council of the London University. He was very active in the Diffusion Society, close to Ricardo, interested in legal and penal reform, and corresponded with Chadwick on health and sanitation.[70] He was probably a central figure at the RI as well. When John Millington resigned his Professorship at the RI, George Birkbeck, who had a particular candidate in mind for the post, saw fit to address that recommendation not to Brande or Faraday, but to Goldsmid.[71]

From the above statistics and portraits, then, we obtain a picture of an institution motivated by the most up-to-date concerns of the English professional class, one at which the notions of science as leisure or even commodity were becoming less influential than what I have described as a legal ideology of science. We could draw on other portraits to demonstrate how remarkable the RI governors were for the time: chemist J. F. Daniell, who invented the constant current battery;[72] the first President of the Statistical Society, the Marquis of Lansdowne; Lord John Russell, the great Whig reformer; the immensely catholic Duke of Somerset, who was RI President during 1827–41; or Henry Warburton, timber-merchant, turned, Philosophical-Radical, who became one of England's great crusaders in the field of medical reform. I am not trying to argue that the RI was wilfully seized by a Benthamite 'conspiracy', and science deliberately bent to some theoretical programme. Rather, what we have in the governorship is a large proportion of men possessing a common pattern to their

---

[68] Robert Edward Anderson, 'Leonard Horner', *DNB*, XXVII (1891), 371–2.

[69] L. T. C. Rolt, *Isambard Kingdom Brunel* (Harmondsworth: Penguin Books, 1970).

[70] *See* sources cited in footnote 38 above; Grobel, *Society*; University College, London, Chadwick Papers; and Claude Joseph Goldsmid Montefiore, 'Sir Isaac Lyon Goldsmid,' *DNB*, XXII (1890), 83–4.

[71] Letter dated 25 October 1829, and contained in the second of two volumes of Goldsmid letters, in possession of University College, London.

[72] '[I]ts action,' he wrote to Charles Babbage, 'even in its present State, had inspired me with the notion . . . of applying it as a Motive power (what Philosopher is it who says that there is 6/8 at the bottom of everything?)' *See* BM Add. MS. 37190, ff. 274–5.

14a. R. I. Murchison

14b. Davies Gilbert

14c. The Marquis of Lansdowne

14d. J. A. Paris

lives that falls under a Utilitarian rubric. These men saw science as a professional tool: the basis for their own expertise and the key to an organized and efficiently administered society. Science in their view was not polite knowledge for an hereditary elite, nor useful knowledge for agricultural improvers (though these were not ruled out), but more importantly a general instrument for creating a smoothly functioning social order. It was the basis for the upgrading of the medical profession and the crux of a new type of legal expertise. Through it, public health might be promoted, forgery prevented, education diffused, pollution curbed, and a modern face for London—bridges, tunnels, housing—constructed. Certainly the Royal Institution was not the Utilitarians' central focus—indeed part of their talent lay in the diffusion of efforts—but their interest in the governing board of such a scientific institution was not accidental, and their impact on policy, procedure, and personnel at the RI hardly minimal. It was, as during the first decade, through the Laboratory and the Professor of Chemistry that the politics of the governors received their most coherent expression; but before turning to Brande (and later, Faraday) it will be instructive to examine the ways in which the overall organization of the Institution itself was influenced by the Utilitarian frame of mind.

By 1825 the Utilitarians had achieved a plurality amongst the governorship, replacing the landed interest's traditional monopoly of the elected offices. The following year the Managers selected as their first choice to replace their late President, Thomas Pelham (Earl of Chichester), the Marquis of Lansdowne, a leading Benthamite and later the first President of the Statistical Society.[73] Between 1826 and 1827, the response to the Friday Evening Discourses (*see* below) brought in 129 new Members, of whom at least a third were in the professional category.[74] The late 1820s also saw a change in the atmosphere of the RI to a more streamlined and business-like sort of efficiency, as well as a general review of its projects and policies—the journal, the lectures, the Institution's financial basis—which reflected a managerial mentality.

Some of the post-1825 changes were subtle, yet indicative of a new type of attitude toward scientific organization. Albemarle Street was macadamized in 1827; the livery outfits of the porters were replaced by plain blue coats and waistcoats, and their duties specifically written down; ballot boxes were instituted at general meetings; and expensive coffee, tea, and gilt-edged stationery were replaced by cheaper and more practical substitutes.[75] The governors began to feel that the lectures should be as scientific as the market could bear. In 1828, the President wrote to the Secretary of his wish to get rid of a planned series of lectures on hieroglyphics. It 'belongs more properly to the Antiquarians & the Dilettanti than to us,' he declared, and the Secretary concurred.[76]

[73] *MM*, VII, 91. Lansdowne, however, declined, and the Duke of Somerset was elected.
[74] Lawyers alone amounted to 24%.
[75] *MM*, VII, 120, 122, 244ff., and VIII, 38–54. *See also* VIII, 394ff. on the streamlining of the Library.
[76] RI Archives, Guard Book Vol. 3, letter from the Duke of Somerset to E. R. Daniell, 1 January 1828. It was by then too late to cancel the series, however, so it was held nonetheless.

The detailed review of the lecture programme that began in 1825 was accompanied by a 'system of retrenchment' and the appointment of several committees to monitor the Institution's finances. This simultaneous development is no surprise: the lectures were the mainstay of the Institution's support. As Joseph Agassi has pointed out, prior to the foundation of the RI one could not attend lectures on science anywhere except at one's local philosophical society; and with the new demand for scientific knowledge, there was clearly money to be made in such a venture.[77] In itself, this was not enough: at periodic intervals the governors had to lend (or give) the Institution very large sums of money, and thus it failed, during the first four decades, to become self-supporting. But the one activity that did bring in a large regular income was the lecture programme, principally in the form of Annual Subscriptions. Annual Subscribers ultimately paid five guineas per year to attend the public courses of afternoon lectures, and in 1826 a review of the programme estimated that the average annual income from such subscriptions during 1823–5 amounted to £1,200—more than 50% of the Institution's total income. Although the figures vary from year to year, this was roughly the situation down to 1826.

Not counting the Friday Evening Discourses, and the Juvenile Lectures held at Christmas, the Institution was, from 1813, sponsoring two types of lecture series, private and public. Public lectures were held in the afternoon, and were on subjects of popular interest, including sculpture, oratory, painting, and other non-scientific topics, as well as on broad areas of science itself.[78] These latter were given mainly by RI personnel. Thus Brande regularly gave a general chemistry course, and John Millington a course in natural philosophy or mechanics; Roget would lecture on comparative physiology; and later Faraday would lecture on chemistry or electricity. Attendance figures are not available prior to 1830, but from 1830 to 1840 both Brande and Faraday experienced a generally steady increase in average enrollment, as the following table shows:[79]

	1830	*1831*	*1832*	*1833*	*1834*	*1835*
Brande	145	167	—	197	200	250
Faraday	—	164	256	247	240	319

	1836	*1837*	*1838*	*1839*	*1840*
Brande	286	201	—	274	283
Faraday	374	425	495	436	412

Faraday, as we can see, was having good success in attracting an audience, and attendance at Brande's lectures was certainly respectable. As for the private lectures, i.e., the morning classes held for medical certification (discussed below),

[77] Joseph Agassi, *Faraday as a Natural Philosopher* (Chicago: University Press, 1971), p. 12.
[78] For lists of lectures and syllabi *see* RI Archives, Vol. 23, Index to the Managers' Minutes, and the collection of cards in Box File V.
[79] Constructed from RI Archives, Box File VII, Item 107.

Members could attend, but not Annual Subscribers. The bulk of the audience, however, were medical students paying four guineas per academic year, and both Brande and Faraday drew roughly the same attendance during the 1830s, about 100 to 120.

The decision to review Institutional policy led to the appointment of a Committee of Lectures in 1825, which gave its report early in 1826.[80] It recommended that the morning lectures be held in the Theatre instead of the Laboratory, so as to encourage greater publicity and attendance, and that private lectures be given on other subjects in addition to chemistry, which it believed would induce Annual Subscribers to become Members, and enable them to attend. It also voted to allow individuals to subscribe to particular courses, which would serve as an introduction to the RI and thus perhaps induce them to become Annual Subscribers. Most of this was done, and it is noteworthy that the two private lecture series decided upon were political economy—'instituted in honour of the late David Ricardo'—and medical jurisprudence.[81] The first choice is interesting, especially since an 1815 offer of such lectures was rejected as being outside the scope of the Institution's interests.[82] J. R. McCulloch, a leading Utilitarian and statistician, whose *Principles of Political Economy* had just rolled off the press, was contracted to deliver twenty-five to thirty lectures on the subject. The second set of private lectures were given by Dr John Gordon Smith, medical and political reformer who became London University's first Professor of Medical Jurisprudence in 1829. Smith stressed the contemporary urgency of the study of forensic medicine and medical police work, and the course included lectures on homicide; marriage and population; regulations for the purity of air, food, and drink; institutions of confinement (prisons, asylums, almshouses, hospitals) as regards ventilation, labour, and discipline; and dangerous situations, such as occur in mining.[83] The Managers also decided to allow separate subscriptions for two of the afternoon courses (two guineas for men, one for ladies); and Faraday offered to give Members weekly demonstrations on chemical subjects.[84] This had been done a few times in the Laboratory during 1825 on an informal basis, and now became regularly organized into the famous Friday Evening Discourses. In May of 1826 the Managers appointed committees to arrange the Discourses and other lectures, and considered reducing the cost of Membership in the hope of increasing the number of Members.[85] It is not clear whether this latter step was taken, but the effect of this revamping of the lecture programme was a dramatic increase in the Membership, as the following figures indicate (numbers refer to Members elected during the year cited):[86]

---

[80] On this *see MM*, VII, 52ff.
[81] ibid., p. 64.
[82] ibid., VI, 45 and 70.
[83] *QJS*, XXI (1826), 116–26.
[84] *MM*, VII, 63. The Minutes do not make it clear whether this was Faraday's idea or the Managers'.
[85] ibid., pp. 78–9 and 82.
[86] Constructed from RI Archives, Item 77, Elected Members 1821 to 1845. Note that in a few

1821—14	1826—48	1831—53	1836—63	1841—25
1822— 8	1827—88	1832—35	1837—35	1842—25
1823—11	1828—78	1833—55	1838—51	1843—26
1824— 6	1829—54	1834—61	1839—37	1844—37
1825—14	1830—57	1835—48	1840—32	1845—24

There is a slight drop over the period of Faraday's nervous breakdown (1839–44), but it would be specious to argue that Faraday's popularity as a lecturer was keeping the Institution solvent. The financial records of the RI reveal that this period was a successful one. There was a regular surplus of income, and the income from subscribers to particular lecture series managed, for the first time in the Institution's history, to make a profit in 1843 of £68, which more than doubled the following year. Faraday's absence was thus hardly an obstacle to the Institution's success.[87]

The Friday Evening Discourses proved to be a winning formula for the RI, and the Annual Reports of the Visitors frequently note the 'unabated eagerness' with which Members and their friends flocked to them.[88] On the face of it, the Discourses would seem to be a reversion to the gentleman amateur tradition: where Utilitarianism fits into this vital activity is hardly apparent. The Discourses retained the trappings of the amateur tradition and do so to this day. Faraday defined them as 'meetings of an easy and agreeable nature', after which 'everyone may adjourn for tea and talk.'[89] They will, he wrote in his lecture notes, 'facilitate our object of attracting the world, and making ourselves with science attractive to it'.[90] But the lectures often had nothing to do with science whatsoever, and the after-talk adjournment to the Library typically involved the display of some curiosity, a practice that was so much a part of the hegemonic tradition. On the evening of 26 January 1827, according to the RI journal, 'some specimens of Dried Plants, from Massachusetts, were presented by Mr. Sharpe, of Boston;' and on February 16th Members were invited to contemplate a stuffed Honduras turkey.[91] Finally, a compilation of the Discourses for the first ten years during which such records are available reveals that 62.5% were on pure science or amateur subjects, and only 36.4% on subjects of technology or applied science (there is no information available on 1.1%).[92] If we are to argue that Utilitarianism had begun to displace

---

instances, individuals withdrew or died shortly after becoming Members. Thus the figures for 1826 and 1827 add to 136, rather than the 129 cited earlier in the text. For the purposes of obtaining a meaningful social profile, it seemed reasonable not to include the seven men whose Membership terminated soon after they joined.

[87] *VM*, Annual Reports for 1840–4.

[88] This particular phrase occurs in the *VM*, Annual Report for 1842 (once again, during Faraday's absence).

[89] *LPW*, p. 329.

[90] *BJ*, I, 392.

[91] *QJS*, n.s., I (1827), 210 and 214–15.

[92] It should be noted that in attempting to classify each Discourse, there were a number of borderline cases that could have gone either way, e.g. Faraday's lecture of 5 May 1826 on hydrocarbons. In this and other cases, I have given the amateur category the benefit of the doubt. Thus the figure of 36% represents the lowest possible fraction of lectures on technical subjects.

amateurism, something must be said about what would seem to constitute an obvious contradiction.

The ambiguity over the nature of the Friday Evening Discourses is embedded within the Utilitarian goals themselves. The Utilitarians, after all, stressed the diffusion of knowledge very strongly, and it is no coincidence that the Friday Evening Discourses were instituted during the same year that saw the formation of the London University and the Society for the Diffusion of Useful Knowledge. With its emphasis on the diffusion of knowledge from its earliest days, the RI would *naturally* attract Utilitarians. The question thus arises, how does one distinguish, in the 1820s and 1830s, between rational entertainment, as defined by the aristocracy, and the diffusion of useful knowledge, as defined by the Utilitarians? Where, in terms of something like a lecture programme, does the amateur tradition end and Benthamism begin?

There is at least one obvious difference. The avowed goal of the amateur tradition was entertainment, of the Benthamites social stability. Hence the former tended to pursue knowledge that was almost sensationalist, in the sense of being odd, whereas the latter were making a bid for relevance. But the major difference lay not so much in content or political purpose vis-à-vis another class as in the attempt to foster a sense of social legitimation amongst the Utilitarians themselves. What is striking in the figures given above is not that 62.5% of the Friday Evening Discourses sampled were dilettantish, but that 36.4% were *not*. We are, after all, talking about pre-Victorian and early Victorian England, when the amateur tradition in science was still dominant. The ratio of technical to amateur topics at the RI was roughly 3 : 5, remarkably high for any type of scientific institution during this time. Faraday's own Discourses—more than 100 during 1825–62—which normally drew between 400 and 600 people, were frequently technological and oriented toward public improvement and industrial innovation.[93] Many topics had associations with the RI. Between 1826 and 1833 there were several Discourses on work being done by I. K. Brunel. Faraday also lectured on his own work on optical glass, dry rot prevention (both done for the Admiralty), the Haswell investigation, lighthouse ventilation (done for Trinity House), etc. In this way, the Discourses served to be self-validating for the new Benthamite group. As Arnold Thackray wrote of the Mancunian middle class, science 'offered a coherent explanatory scheme for the unprecedented, change-oriented society in which [social reformers] found themselves unavoidably if willingly cast in leading roles'. We see here 'the ability of science to function as ratifier of a new world order'.[94] There was a dual message being delivered at these meetings that was every bit as ideological as Davy's praise of the agricultural improvers: (1) science has everything under control, and (2) it is *you*, this particular group of people, who hold this tool in your very hands. In the chaos of the Industrial

---

[93] S. P. Thompson, *Michael Faraday* (London: Cassell, 1898), p. 225; *LPW*, pp. 331–2.
[94] Thackray, 'Natural Knowledge in Cultural Context', pp. 682 and 686.

Revolution, Faraday was presenting a comprehensible world, order and stability, with his audience pictured as being in the driver's seat. The Discourses were thus often a tribute to Benthamism, although it would be stretching a point to regard this trend as being particularly deliberate. Faraday was no Utilitarian; but he had certainly found an audience. He 'showed how science could react upon a world largely indifferent to it', writes Williams;[95] but the Members of the RI, and most especially the governors, were not so indifferent. In the Friday Evening Discourses, Whigs and Benthamites, improving physicians and reforming barristers, were able to see themselves reflected both as polished gentlemen *and* as men leading the nation toward social stability and modernity. There was no reason, furthermore, why these lectures, which were so important to the Institution's finances, could not be entertaining, politically stabilizing, and socially legitimating, all at once.

The financial history of the RI is also a story of the successful combination of Utilitarian and aristocratic factors. We see the former in the constant attempt to render the Institution efficient and self-supporting. But ultimately the RI never (during 1799–1844) escaped the tradition of aristocratic patronage, and as non-professional assumptions affected its development considerably, it might be useful to review its financial career from the Davy years down to the mid-forties.

The entrepreneurial vision which Davy advertised did not, in his eyes, apply to science itself, a viewpoint that was in the mainstream of British thinking about the subject. That is to say, Davy, and the early governors, generated much enthusiasm over the economic potential of science, but never saw science *itself* as an enterprise. In refusing to patent inventions, Davy was setting himself squarely within the hegemonic tradition; and although there were to be some important exceptions (such as Lord Kelvin), for most of the nineteenth century payment for scientific work was regarded as bourgeois, if not even vaguely corrupt. How then was a scientific institution to support itself? This was not much of a problem prior to the foundation of the RI, since such organizations that did exist were supported on a voluntary basis (although the Royal Society did receive occasional subsidy for specific projects). But the RI was to be a going concern, and its annual expenses were heavy. Could friendly patronage be a reasonable *modus operandi*?

In the absence of alternative models of scientific support, the answer had to be yes. Davy's untiring campaign for support, which went on during all of his adult life, was for *more* patronage, not for a different type of support. The professionalization of science—income derived from activity, practitioners as paid employees—was antithetical to his goals and accounts for much of his resentment towards Brande (*see* below). His idea of remuneration was largely one of rewards and incentives, and as President of the Royal Society he obtained concessions such as the establishment of the Royal Medals, the foundation of a zoological park, and the granting of additional apartments for the

[95] *LPW*, p. 332.

Royal Society.[96] But all of this was easily accommodated within the amateur tradition. Davy was, in reality, a hybrid, a 'professional amateur' who simply wanted more support for the virtuoso tradition on the part of wealthy patrons.

There was, however, a significant drawback as far as the RI was concerned: neither the older formula of patronage, nor the newer one of corporate continuance, really worked, at least not by themselves. The RI spent its first four decades in debt, and came very near to closing on more than one occasion. The first crisis arose in 1803, when the governors saved the Institution by lending it £2,000. During 1799–1802 the principal source of income—nearly £18,000—was from Proprietorships, but the expenses of the house itself exceeded £19,000, and there were few new Proprietors after 1803.[97] From 1803 to 1826, people tended to join the RI on an annual basis, and between 1803 and 1814 the Annual Subscriptions amounted to roughly £1,800 per annum. But the upkeep and maintenance of the house (there were, for example, nearly thirty people on its staff) averaged £1,700 per year during 1803–14, and the lecturers' fees came to £900. The governors once again loaned the corporation £2,300 in 1808, but its debts were over £2,000 by 1809, and £4,000 worth of consolidated annuities, or 'Consols', as they were called, had to be sold. By 1814, the debt was up to £1,924, and the RI was rescued by an emergency £1,815 raised from Members and Life Subscribers.[98] The Institution was, however, soon in debt again, and struggled with a large deficit—£2,395 by the end of 1822—until 1823, when a number of governors and Members provided loans amounting to £3,700. Over the next few years, these loans were gradually converted into gifts; and this endowment, plus the success of the Friday Evening Discourses, pulled the RI out of debt by 1836–7.

What was the reason for this nearly endless debt? An examination of the annual expenses over the years does not reveal any unusual items, but the house itself—salaries, lectures, upkeep, the whole operation—typically ran to £3,300 per annum, and the money which the lectures earned in the form of Subscriptions or Memberships was inevitably much less. Herein lies a crucial irony: the amateur tradition dictated that only the public lectures earn money. From Brande's medical lectures, for example, the Royal Institution took only one third of the subscription, and this exceeded £100 only once, in 1830. From the thousands of analyses performed over the years, the RI took nothing (!). Despite the generally accepted belief that this work kept the Institution solvent, the financial records of the Laboratory do not bear this out.[99] The RI paid for all the apparatus, chemicals, and other expenses of the Laboratory, and yet

[96] *DW*, VII, 57; Fullmer, 'Davy's Sketches of his Contemporaries', *Chymia*, XII (1967), 127–8; Davy to Sir John Leslie, 21 January 1824, in the collection of W. L. Marjoribanks of Aberdeen; Davy to Thomas Bernard, 8 December 1816, in RI Archives (purchased from John Wilson on 13 May 1974).
[97] *See BJ*, p. 425. By 1807 there were 374 Proprietors, and an upper limit was set at 400, with the cost now 200 guineas. *See also VM*, Annual Report for 1807.
[98] RI Archives, Report of the Managers for 1809, Box File XIV, Folder 131; *VM*, Annual Reports for 1813 and 1814.
[99] Cf. *LPW*, pp. 322 and 359n; and the Laboratory accounts in the *VM*.

regarded the commercial work done by its scientists as none of its affair. Brande, and presumably Davy, pocketed their income; Faraday gave most of his to charity. This was, in fact, laissez-faire with a vengeance. In 1830 Faraday earned £1,000, and Silvanus Thompson estimated that he could have made £5,000 per annum if he chose. '. . . I can at any moment convert my time into money,' Faraday wrote to Trinity House in 1836.[100] In short, the RI *could* have paid its way and have even operated at a profit: instead, because of the amateur tradition, it nearly went bankrupt several times.

Nevertheless, the amateur tradition did not leave the Institution stranded: it was patronage that saved it and patronage that made its evolution to the status of corporate self-sufficiency possible. True, the Friday Evening Discourses, with their technological orientation, were crucial to this evolution, but a number of large gifts were also essential. During his lifetime John Fuller gave the RI more than £10,000; the outstanding Library—more than 20,000 volumes by 1820—was paid for by voluntary subscriptions; the Laboratory fund was supplied by a number of random donations, as well as by a voluntary one-guinea-per-year contribution made by many of the Members; bequests came in from wills during 1825–37; and the creditors of 1823 ultimately decided not to reclaim their £3,700 loan. By 1840, the Institution actually had a surplus of £400, and had managed to accumulate roughly £10,000 in Consols.[101]

In terms of organization, then, the RI proceeded from 1811 to 1840 by a judicious mixture of amateur and professional elements, and thus managed to survive. But with the exception of the lecture programmes, the projects undertaken by the Institution during the years of Brande's tenure were unequivocally Utilitarian in nature, and it was this that attenuated the amateur tradition most of all. In the person of William Thomas Brande the Utilitarians found a professional technician, the concrete embodiment of their goals. The years dominated by Brande's leadership (ca. 1813–31) seem very pedestrian indeed in comparison to the first decade at the RI. Yet during this same period, Brande was commonly regarded as London's leading chemist. The revived RI journal, to take one example, was matter-of-factly called 'Brande's journal'. If Brande's lacklustre diligence has failed to find him a single biographer, it is nevertheless the case that nearly two decades of the RI's history, as well as its subsequent development under Faraday, were decisively shaped by Brande's medical, commercial and strongly professional concerns.[102]

Brande was an apothecary by blood as well as by temperament. He came

[100] Thompson, *Michael Faraday*, pp. 63 and 66–7.
[101] This information is taken from the Annual Reports in the *VM*; some of it is reproduced in the *MM* as well.
[102] The only extended study of Brande is by Aubrey A. Tulley, *The Chemical Studies of William Thomas Brande 1788–1866* (M.Sc. Dissertation, University of London, 1971). Short sketches of Brande include Robert Hunt, 'William Thomas Brande', *DNB*, VI (1866), 216–18; E. L. Scott's piece in C. C. Gillispie (ed.), *Dictionary of Scientific Biography*, II (1970), 420–1; Edward Ironmonger, 'Forgotten Worthies of the Royal Institution: William Thomas Brande (1788–1866), *Proceedings of the Royal Institution of Great Britain*, XXXVIII (1961), 450–61; and C. H. Spiers, 'William Thomas Brande, Leather Expert', *Annals of Science*, XXV (1969), 179–201. The portrait of Brande that follows is based on these sources.

from an apothecaries' family, German by birth, whose members had served as medical advisers to royalty. Between 1741 and 1835 the Brande family ran an establishment in Arlington Street, where Brande was born in 1788. Brother, father, grandfather, and great uncle were all associated with it, and Brande himself began his apprenticeship in 1802. From that point, his career followed a natural and unwavering course, but took on a strongly professional bent due to the agitation for the Apothecaries' Act that began in 1804. He studied with Charles Hatchett (his future father-in-law), and formed a society for the study of animal chemistry with Hatchett, Davy, and two physicians, Everard Home and Benjamin Brodie, in 1808. Brande also studied at the Windmill Street medical school and at St George's Hospital, and shortly after completing this training returned to both in the capacity of an instructor. In 1808 he lectured on pharmaceutical chemistry at the Cork Street medical school, on physics and chemistry in Windmill Street, and gave private lectures on *materia medica*. Brande published his first article at the age of sixteen, became an F.R.S. at twenty-one (in 1809), and served as RS Secretary, 1816–26. In 1812, when Davy was unable to deliver his spring lecture series to the Board of Agriculture, Brande gave them in his stead, and the same year was made a professor at Apothecaries Hall. Sometime later, true to his interest in giving medicine a chemical basis, Brande created a new post for himself, that of superintending chemical operator at the Hall. Finally, he became the Royal Institution's Professor of Chemistry in 1813 and remained in that position until 1852. As John Davy once remarked, Brande managed to tie down almost all of the key scientific-administrative posts available. In his later years, he became Master of Apothecaries Hall (1851), Chief Officer of Coinage at the Mint (1825–54), and an examiner of London University (1846–58).

Although much of Brande's research, especially his laboratory work at the RI, consisted of commercial analyses, medicine remained the hub of his professional life. Brande saw his goal as one of deepening the relationship between chemistry and medicine.[103] Most of the articles he published during his lifetime dealt with some aspect of that grey area between physiology and organic chemistry. From 1805, a steady stream of pamphlets and articles on respiration, calculi, chloroform, uric acid, animal fluids, wax, blood, tea, starch, ether and urine flowed from his pen. Two of Brande's books—*A Manual of Pharmacy* (1825) and *Dictionary of Materia Medica and Practical Pharmacy* (1839)—became standard reference works and popular texts in medical education. The *Manual*, in fact, was based on Brande's lecture course at Apothecaries Hall, and the *Dictionary* was dedicated 'to the students of the metropolitan medical schools'.[104] Both at the Hall and the RI, Brande's contacts with physicians and medical students were close, and he soon became nearly indispensable to the new medical community. In so doing, he created a

[103] Tulley, *Brande*, p. 113, and *see also* p. 39.
[104] W. T. Brande, *A Dictionary of Materia Medica and Practical Pharmacy* (London: J. W. Parker, 1839). On the *Manual* see Tulley, *Brande*, p. 74.

very special niche for the Royal Institution. From 1813, most of his medical analyses took place in its Laboratory, and were publicized in its journal; and the Institution also came to serve as a place of medical instruction and qualification. Yet, as we shall see, Brande emerged by the mid-1820s as someone much larger than a medical man or scientific apothecary. Brande's emphasis on scientific expertise, which manifested itself in the codification of technical knowledge, strongly caught the attention of the professional community, and the RI was publicly identified with this work.

It is interesting that Brande fulfilled Davy's promise of the public importance of science beyond anything Davy had ever imagined possible, for no two men could have been more unlike, or disliked one another more; and this difference of personality can be taken as a metaphor for the ideological changes that began to occur after the loosening of electoral policy during 1809–11. Davy was the perfect mirror of the improving landlords. He combined the traditional aristocratic conception of science with the avant-garde agricultural interest in profit and estate exploitation. As we have seen, his actual role was that of an ideologue—perhaps demagogue—and the impact of his propaganda far exceeded any results that could be measured in terms of cost-benefit analysis. His whole notion of scientific organization was aristocratic, however: at best a system of patronage rather than of team research or meritocracy paid for by institutional profits or investments. His personal goals were purely intellectual and even romantic as regards the natural world.

With the passing of the agricultural improvers from the scene, and the influx of the professional classes, Davy's public style became more and more superfluous. We have seen that the motive behind the changes of 1809–11 was an attempt to get away from a patronage and endowment framework that could not, or so the governors believed, secure the RI's financial future. The Institution had to become something of a business venture if it were to survive, and in Brande the governors found someone with a business-like turn of mind. Brande was, quite simply, the scientific version of a Dickensian Gradgrind, and under his leadership public advertisement was augmented by a good deal of private drudgery. Despite (or because of) this, Brande may have done more to shape the Institution's future than either Davy or Faraday.

The lack of interest in chemistry as the investigation of nature, and the deliberate commercial and medical orientation that Brande brought about, was something Davy could not forgive. He once referred to Brande as 'a very inferior person [who] followed chemistry as . . . a metier for small distinctions & as much profit as He could obtain'.[105] John Davy wrote of Davy's view of him:

> Mr Brande very much disappointed my Brother; as a young man he thought well of him; in climbing he shewed much activity—and in a few years he had possession of almost every appointment of emolument within his reach; at the same time he was Professor of chemistry to the Royal Institution, Professor of materia medica to the Company of Apothecaries, inspector of drugs to the East India Company,

[105] Fullmer, 'Davy's Sketches of his Contemporaries', p. 134.

superintendant [*sic*] at the [M]int, Secretary to the Royal Society & Editor of a Journal.—He was mercenary and had no lofty views;—he had come from a counter, his father was an apothecary,—& he was rather fitted for it than for a Professors or Secretary's chair.—During the many years that he has had the direction of the Laboratory of the Royal Institution he has not made a single discovery or contributed any thing to the advance of chemical science.[106]

'He had come from a counter . . . .' It was this that rankled Davy, to whom Brande, for his part, contemptuously referred as 'that self-constituted autocrat of science . . .'.[107] Yet the fact that an apothecary could rise, in a few short years, to the directorship of the Royal Institution says more about the social changes currently taking place than it does about the talent of the man himself. If Davy was a reflection of the improving landlords, Brande was the epitome of the Utilitarian interest, and the fact that he had no London equal in the 1820s bears witness to what he had become: the Bentham of administrative chemistry. Echoing Bazarov, he would have agreed with the latter's proclamation in *Fathers and Sons:* 'Science is a workshop, not a temple'.[108]

The shift that had taken place at the RI was captured somewhat caustically by Thomas Carlyle, who deeply resented the passing of an organic society in favour of a coordinated one. In an essay written in 1829, which he pointedly entitled 'Signs of the Times,' Carlyle complained that

[n]o Newton, by silent meditation, now discovers his system of the world from the falling of an apple; but some quite other than Newton stands in his Museum, his Scientific Institution, and behind whole batteries of retorts, digesters, and galvanic piles, imperatively 'interrogates Nature',—who, however, shows no haste to answer.[109]

Carlyle was wrong, of course, in suggesting that with scientists like Brande, nature 'showed no haste to answer', for Brande was asking different questions. And it was hardly the case that the age of the independent scientist had passed. Yet Carlyle did capture the flavour of the RI under Brande, the conversion of science into a business and (as much as was then possible) a profession. Davy attempted to resist this trend, but by 1820 his influence at the RI was considerably diminished. At a meeting he attended in 1816 the Managers passed a resolution that 'due attention [be given] to preserving the Laboratory as a place of Experimental research on Chemistry',[110] but this was clearly a courtesy to the former Professor of Chemistry, and had little effect. Despite the stupendous discoveries made by Faraday over the years, for example, the Managers'

[106] RI Archives, Davy MSS, Vol. 14, notebook j, pp. 165–7.

[107] Samuel Smiles, *A Publisher and His Friends: Memoir and Correspondence of the Late John Murray* (2 vols.; London: John Murray, 1891), II, 208.

[108] The exact quotation in the New American Library translation (1961), is: 'Nature is no temple but merely a workshop, and man is the craftsman' (Ch. 9).

[109] [Thomas Carlyle], 'Signs of the Times', *The Edinburgh Review*, XLIX (1829), 443. Identification of the author is in Louis Cazamian, *The Social Novel in England 1830–1850*, trans. Martin Fido (London: Routledge and Kegan Paul, 1973; orig. ed. 1903), p. 34.

[110] *MM*, VI, 81.

Minutes never bothered even to mention them in their pages.[111]

Brande's first move in the direction of professional service was, as might be expected, in the field of medicine. Soon after his arrival at the RI, he proposed using the Institution's laboratory as a classroom for medical students, and with the passage of the Apothecaries' Act the Managers agreed

> that it would be highly advantageous as well as hono[u]rable to the Institution, to give authority to M͟ Professor Brande to transfer his Course of Lectures on Chemistry from the room in Windmill Street to the laboratory of the Royal Institution . . . .[112]

The reasons for such a move are quite clear. The Apothecaries' Act (now under consideration for several years) specified courses in chemistry and other branches of science as a prerequisite for qualification. The student had to attend a series of lectures, pass an examination, and present a certificate testifying to this work to Apothecaries Hall. By transferring his lectures—and examinations[113]—to the RI, Brande was giving the Institution the role of catering to an important sector of the new professional class. Would-be physicians, surgeons, and apothecaries who studied at the Windmill Street school and St George's Hospital were soon seen hurrying down Albemarle Street thrice weekly for their 9 a.m. lecture. 'How often we have stood,' wrote Faraday's niece, who lived for a time at the RI,

> at the top of that long well [*sic*] staircase—& counted the pupils as they trooped down to the 9. o'clock lectures! I think there were generally between 90 and 100 of them. In the winter time it was hard work to be ready to begin lecturing at 9.—Very quiet was the breakfast on those mornings, three in the week, Tuesday Thursday & Saturday and very little breakfast did dear uncle eat or indeed any of us.[114]

Faraday took this responsibility very seriously, having his friend Edward Magrath attend the lectures so as to advise him on his delivery; and as time went on, Brande relied on his services more and more.[115] In 1827–8, Faraday gave eleven out of the thirty-four lectures, and in 1835 added a course of lectures on electricity for these students. This was an excellent example of the use

---

[111] And this includes the discovery of electromagnetic induction in 1831. The first reference to Faraday's pure researches was in 1845 (rotation of the plane of polarized light in a magnetic field), and it was on Faraday's initiative (*MM*, IX, 357).

[112] *MM*, V, 394, and VI, 77. This was tried on a very small scale as early as 1811, when medical students were admitted to Davy's lectures at two guineas per course. *See MM*, V, 250–1, 265, 275, and 278.

[113] Reference to course examinations is made in a letter to the Managers from Thomas Griffiths, who studied with Brande and Faraday and took over for Brande in cases of illness or absence, and conducted the examinations on several occasions. *See* RI Archives, Box 16 (old numbering system), Item 179, dated 19 July 1852.

[114] RI Archives, Diary of Margery Reid, p. 157. The letters 'a.m.' appear over the word 'o'clock' in the original. Margery Reid was the niece of Sarah Faraday.

[115] ibid., pp. 163–4. Faraday began sharing the lectures with Brande as of 1825; *see QJS*, XVIII (1825), 199.

of science for the manufacture of professional credentials, rather than for the acquisition of actual medical skill, as one student, Bence Jones, recognized:

> On October 1st, 1838, I became a perpetual pupil at St. George's and began to attend lectures and the dissecting room. I then for the first time went to the Royal Institution, where Mr. Faraday, at eight in the morning, gave a short course of lectures on electricity, which preceded Mr. Brand's [*sic*] general chemical course for the medical students of St. George's.
>
> Those lectures of Mr. Faraday's, beautiful as they were, were of small use to medical pupils, who could not see much connection between electrical induction and the action of drugs.[116]

We would not have a record of the 9 a.m. lectures had it not been for Thomas Wakley's opposition to monopoly in the medical profession. The *Lancet* editor was a close friend of one RI governor, Henry Warburton, Parliamentary champion of medical reform; but he was an outspoken opponent of 'Rhubarb Hall', as he dubbed the Apothecaries Society, and did not like the idea of a private series of paid lectures.[117] As a result, *The Lancet* undertook an extensive and possibly verbatim coverage of Brande's lectures, and it was very flattering publicity. Wakley cited the RI in his column, 'The Medical and Surgical Schools of London', and described the course as including 'a full examination of pharmaceutical chemistry; the processes of the pharmacopoeiae will be similarly described and compared with those adopted by manufacturers'.[118] If Wakley opposed monopoly control, it was nevertheless the case that he thought Brande's lectures first-rate, and was as much a proponent of chemistry as the basis of medicine as was Brande. 'The medical man,' he wrote in a review of Brande's journal, '*ought* to be a scientific man.'[119]

Wakley's transcription of these lectures, however, reveals nothing extraordinary in them. All of the information which they contained required no original research on the part of the lecturer, could be compiled from any number of respectable secondary sources, and dealt with topics such as heat and affinity at the level of a survey course. It was, however, Brande who compiled them, the RI that hosted them, and through which the information was made neatly accessible. Brande was not being immodest when he stated, upon his resignation in 1852, that they 'were the first lectures in London in which so extended a view of chemistry, and of its applications, including technical, mineralogical, geological, and medical chemistry, was attempted ...'.[120] The value of the lec-

---

[116] Henry Bence Jones, *An Autobiography* (London: Crusha and Son, 1929), p. 13.

[117] Charles Brooke, *Thomas Wakley* (London: The Socialist Medical Association, 1962), p. 15; *The Lancet*, IX (1925), 3–4, and the issue for 6 October 1827, p. 3. Brande did an analysis of rhubarb in 1821.

[118] *See* issues beginning with 20 October 1827 and also the issue for 26 September 1829, pp. 11–12. The lectures began to attract more attention when the Managers had Brande shift them from the Laboratory to the Theatre (*MM*, VII, 52ff.).

[119] *The Lancet*, II (1828–9), 73.

[120] 'Resignation of Professor Brande', *Notices of the Proceedings at the Meetings of the Members of the Royal Institution*, I (1852), 168. *See also MM*, IX, 416; Leslie G. Matthews,

tures for the actual practice of medicine was minimal, but this made little difference. The belief in scientific expertise and its relevance for professional standards was crucial for the Institution's reputation and development. Through the lectures the RI was becoming publicly identified as a service organization for the medical community. As the Managers themselves noted, the lectures were 'productive of great advantage to the Royal Instit." and have greatly raised its character as a school of Chemistry . . .'.[121]

Two books crucial to medical education emerged from the Brande-Faraday lectures and quickly became popular texts. According to Brande, his *Manual of Chemistry* (1819) was originally intended as a text for the chemistry students at the RI and was written, with Faraday's help, partly in the Laboratory itself.[122] As a handbook of chemical knowledge, an orderly compilation of what Brande might have called 'known and useful information', it was unequalled. It was heavily documented, and meant to be used as a working guide. In 1828, when a Select Committee of the House of Commons was interviewing a physician (and Member of the RI), Robert Kerrison, on the subject of the London water supply and the problems of pollution, Kerrison quoted from Brande's *Manual* at length. Both he and the Commissioners took it to be the best all-round guide to theoretical and practical chemistry.[123] The second book, Faraday's *Chemical Manipulation* (1827), surprisingly the only book he ever wrote, was designed to meet the needs of students seeking to acquire sound laboratory technique. It should be added that the current educational regulations of the Apothecaries Company stressed the importance of 'instruction in chemical manipulation and practical analysis . . .'.[124]

It was thus through lectures, laboratory analyses, and textbooks that the RI, via Brande and his young assistant, was drawn into medical professionalization. By being so absorbed, moreover, it managed to make its own continuation possible. In finding a role for science in the new professional wave, Brande laid the groundwork for the professionalization of science itself. Science did not become professionalized in terms of its own content, especially since research was so closely associated with the amateur tradition. To the extent that the RI provided a model, we can say that science became professionalized in terms of its ability to accelerate the process of professionalization in other spheres.

History of Pharmacy in Britain (London: E. and S. Livingstone, 1962), pp. 116 and 158; Edward Ironmonger, 'The Royal Institution and the Teaching of Science in the Nineteenth Century', *Proceedings of the Royal Institution of Great Britain*, XXXVII (1958), 144 and 147.

[121] *MM*, VII, 54. The RI made relatively little in terms of profit, as the courses cost four guineas each but Brande was allowed to keep two-thirds of the total, and the RI bore all the expenses of the lectures. See *MM*, VI, 81, and *The Lancet* for 26 September 1829, pp. 11–12.

[122] *See* the preface to the 1848 edition, and Tulley, *Brande*, p. 49. Faraday bound a copy for himself, alternating the pages with blank interleaves, on which he made copious notes. This is preserved in the Library of the Wellcome Institute of the History of Medicine.

[123] Report from the Select Committee on The Supply of Water to The Metropolis. Ordered, by the House of Commons, to be Printed, 19 July 1828.

[124] *Regulations to be Observed by Students intending to qualify themselves to practice as Apothecaries, in England and Wales* (London: Gilbert and Rivington, 1833), p. 7. These were probably drawn up by Brande.

15a. Edward Magrath

15b. Sir John Fuller

16. W. T. Brande

17a. Henry Bence Jones

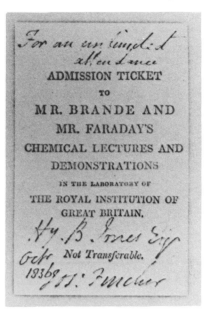

17b. Bence Jones' ticket of admission to the early moring medical lectures

18a. A certificate to be used as part of a student's medical qualification, signed by Faraday

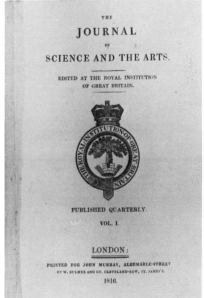

18b. Title page of Vol. 1 of Brande's Journal

19. The Laboratory of the Royal Institution

20. The Laboratory of the Royal Institution

The second contribution toward medical professionalism sponsored by the RI was the publication of a journal (the *QJS*) under Institution auspices, which ran from 1816 to 1831.[125] Out of financial necessity it preserved an interest in the traditional concerns of amateur scientific research. Nearly half the articles were devoted to topics such as meteorology, natural history, ancient mathematics, and so on, and in this way it bore a certain similarity to the *Philosophical Transactions*. But the differences between it and the *Phil. Trans.*, or the publications of David Brewster and William Nicholson, were apparent from casual perusal. It was strongly medical in its orientation, with 93 out of a total of about 816 articles devoted to medical subjects proper, and many of the botany or natural history pieces having deliberate relevance to pharmacology and *materia medica*.[126] In addition, roughly thirty-nine articles were on matters of Utilitarian concern, such as population, vital statistics, street illumination, water pollution, insurance tables, and prisons; and twenty-eight dealt with industrial applications of science. Secondly, the journal was intended to be a working guide, serving to keep the technical reader up-to-date, and was thus a reflection of Brande's handbook/textbook style of organizing factual information to meet modern needs. Brande saw it as a tool rather than, as with the *Phil. Trans.*, an archive. It also became the logical place to turn for such information, and Brande at one point made mention of 'the numerous queries that are put to us' respecting a variety of technical issues.[127] In fact, for a pre-Victorian publication it may have been *too* narrowly conceived. A committee set up in 1830, to work out the details of editing a new journal, remarked that prior to 1827 Brande's journal had been 'principally addressed to the proficient';[128] and the issues from 1827 onwards tended to be more amateur in content. The overall impact, however, was the new emphasis on the medical, professional, and Utilitarian aspects of science.

All of this, of course, came to be commonly associated with the RI. The journal gave the RI the appearance of a technical clearing-house for a society in the midst of modernization. Like its predecessor, which Davy and Young had edited for two years, Brande's journal carried the Institution's seal and reported on RI events; but whereas Davy largely abstracted technical information from other periodicals (usually French), Brande was frequently reporting on analyses done at the RI itself. It thus had the appearance of a 'message from the front'. In general close attention was given to Institution activities, and Brande used it to build the Institution's 'active' reputation. Thus the 9 a.m. lecture series promised an inquiry 'into the application of Chemical principles to the treatment of Diseases', as well as the 'applications of Chemistry to the Arts and Manufactures, and to economical purposes . . .'.[129] Brande's article 'On the Ad-

---

[125] *See MM*, VI, 78–9.

[126] By 'articles' I do not mean reviews, abstracts, and miscellaneous reports, which were often on medicine as well.

[127] *QJS*, XVIII (1825), 1.

[128] *MM*, VII, 297.

[129] *QJS*, I (1816), 309; cf. II, 213–15.

vancement of Science, as connected with the Rise and Progress of the Royal
Institution', pointed out that the 'agriculturalist, the manufacturer, and the
miner, have all received ... gratuitous and cheerful aid', and added: 'As a
school of Chemistry, we boldly challenge competition'. Referring to the 9 a.m.
lectures, Brande noted that 'the policy of our Colleges requires from the
Medical student certificates of his competence in this necessary knowledge
. . .'.[130] Discussions of RI events and their general value to the community
punctuate the journal at regular intervals, and, true to Brande's commitment,
similar advertisements appear for Apothecaries Hall. Finally, the periodical had
all the appearances of a 'house' journal. There were numerous contributors, but
those that can be called regulars proved to be associated with the Institution in
some way. Besides Brande and Faraday, there was John Newman, the
Institution's official instrument-maker; Andrew Ure, the journal's principal
reviewer; and J. F. Daniell, Richard Phillips, H. T. Colebrooke, Davies Gilbert,
and A. B. Granville, all RI governors. Of the approximately 816 articles
published between 1816 and 1831, these nine men alone wrote 151—nearly
20%—and the figure is probably higher in that the journal began publishing
unsigned articles as of 1827.

Although the governors of the RI found the orientation of the journal to their
liking, Brande's zeal in trying to make it Britain's number one scientific
publication frequently got out of hand. In this, as well as in several other ventures, Brande revealed the nakedly aggressive side of professionalism, for he
saw, in advance of almost any scientific contemporary, an expanding market for
expertise which he was determined to corner for his personal advantage. This
competitiveness came out in a series of letters to his Scottish counterpart, Andrew Ure.[131] 'Let us work hard,' he wrote to Ure in December of 1821, 'to
make the Journal *popular* and we shall soon beat Brewster & his crew out of
the field.'[132] 'Our last Number,' he wrote two months later, 'is universally liked
& has been more generally read than any other similar publication.'[133] One
month later Brande continued in the same vein:

> Our last Number was extremely liked though it created some jealousy and ill feelings
> in certain quarters—but this is in our favour—& the first step towards being read is
> to be abused . . ... If you and I can stick to the work, we shall carry all before us
> . . ..[134]

Finally, later in 1822, Brande wrote to Ure: '[W]hat a wretched farrago of

[130] ibid., III (1817), v and xii–xiv.

[131] These letters are in the RI Archives, Box File IX, Item 122E, and run from 1821 to 1825.
Andrew Ure (1778–1857) made his career as a chemical consultant, compiler of technical
reference works, and scientific administrator, becoming very famous in his time for activity
similar to that of Brande. He was a physician, lecturer, and leading promoter of the idea of
applying chemistry to industry. He served as chemist to the Board of Customs, and was one of
the earliest to establish popular scientific lectures for the working class.

[132] ibid., letter dated December 1821. 'Popular' here refers to circulation, not amateur content.

[133] ibid., letter dated 18 February 1822.

[134] ibid., letter dated 20 March 1822.

shreds and patches was the last number of Brewster—surely we must be gaining ground upon them . . .'.[135] In general, the letters reveal a constant preoccupation with circulation and reputation.

Faraday's participation in the journal was considerable. Brande often left him in charge when he was out of town for any length of time, a practice which began as early as August 1816. Faraday wrote to his friend Benjamin Abbott that 'it has had very much of my time and care, and writing, through it, has been more abundant with me'.[136] This last remark proved to be no exaggeration: during 1816–31 Faraday's contribution of articles, reviews, and reports, signed and unsigned, amounted to 131 separate pieces.[137] Faraday also served as editor of the 'Miscellanea' section, which had something of an orientation towards industrial chemistry. Serious differences arose, however, between Brande and Faraday regarding the publication. For his part, Faraday was oblivious to Utilitarian politics or any politics. He naively regarded science as something that inevitably led to harmony, whereas Brande's interest in professional advancement led him to attack (and create) various opponents. Ure's reviews, in particular, were often caustic, and Faraday was completely alienated by 1822, refusing any type of managerial role. '. . . I have nothing to do with it,' he wrote to Gaspard de la Rive,

> not even with the Miscellania [*sic*] now and I am anxious you should know this. There have been many things in the form or [*sic*] reviews observations etc. that have appeared in it wh[ich] were very adverse to my feelings and I am desirous that ther[e] should not be the least chance of your attributing any [of] the kind to me . . ..[138]

As late as 1832, when the journal was no longer being published, Faraday felt the need to state publicly that he had not had a role in its editorship.[139]

Controversy notwithstanding, the journal enjoyed great prestige and a respectable circulation. Normally 1,000 copies of each issue were printed, and the bound annual volumes ran to a second edition. Financially, however, it just managed to get by. The appointment of Brande as editor included the provision that he was to receive all the profits of publication,[140] but these were so trifling that Brande was in effect doing the job gratis. This was, however, not a problem for him, inasmuch as professional reputation, and not financial reward, had been his objective here. Nor did the Managers see the venture any differently, for when they decided, in 1830, to edit it themselves, remuneration was of no great concern. The 'work is to be no source of profit to the R[oyal] Institution or to the

[135] ibid., letter dated 25 September 1822.

[136] *BJ*, I, 224 and 232.

[137] Alan E. Jeffreys, *Michael Faraday: A List of His Lectures and Published Writings* (London: Chapman and Hall, 1960).

[138] *LPW*, I, 139, and *see also* letter to Alexander Marcet on p. 121. For one of Faraday's strongest objections to polemics in science *see* his letter to Matteucci (1853) in *BJ*, II, 321.

[139] *Phil. Mag.*, XI (1832), 236.

[140] *MM*, VI, 78–9, and VII, 295.

Managers', Faraday wrote to his German colleague Mitscherlich in August of that year.[141] The decision to take the journal out of Brande's hands is thus somewhat puzzling. Faraday's version of the story was that the Managers sought to make the journal 'truly scientific', by which he meant that they would exclude 'illnatured [*sic*] reviews'; but why they would wait until 1830 to decide that Brande was too harsh a critic is unclear.[142] Perhaps the simplest explanation is that the Managers wanted the periodical directly identified with them, rather than with their delegated employee. The committee appointed to review the matter noted that the journal 'is and always has been considered the private property of Mr Brande, and that the Managers have no control whatever over the publication'.[143] As a result, they stated, the content, the form of publication, and the manner of distribution have not been subject to their discretion.[144]

> The Committee cannot ... avoid expressing their feelings that if the Journal had been made more substantially a publication emanating from the Managers of the Royal Institution than it appears to have been considered, and that if more advantage had been taken of the resources of the Royal Institution in the preparation of its different parts, it would have been more productive of advantage to the interests of Science .... And they are of the opinion that the Board of Managers in so completely divesting themselves of the control over the publication by delegating of their authority to an Editor have ... in some measures neglected the interests of the Institution ....[145]

There is no way of knowing what was meant by 'the interests of Science', especially since the resulting *Journal of the Royal Institution of Great Britain* proved to be much the same as Brande's. But the new version was seen as 'emanating from the Managers', not just from Brande, and it now had a group of men acting as its editors. Faraday, in particular, resumed an active role, especially in the capacity of a referee.

Although these reasons may have been sensible, the net result was poor management and a rapid end to the entire experiment. When profits from the first four issues amounted to less than £68, the Publication Committee decided to print fewer copies of No. 5, raise the price from six shillings to 7/6, and pay its contributors less. As a result, the venture collapsed completely. At the last meeting of the Committee, according to their minutes, 'The Question of the probability of finding means to Continue the publication of the Journal was deferred'. In addition Faraday, who had been devoting substantial amounts of time to the project, was becoming less and less willing to do so. The fifth issue appeared, but after that, publication ceased.[146]

[141] *LPW*, I, 181–2.
[142] ibid.
[143] RI Archives, Item 102, Minutes of the Journal Committee, 1830–2, entry for 4 January 1830.
[144] *MM*, VII, 301.
[145] ibid., pp. 296–7.
[146] RI Archives, Item 102, Minutes of the Journal Committee, entries for 23 September 1831 and 16 February 1832.

On balance, the journal can be considered a success, because it paid its own way and lasted a full fifteen years. How Benthamite it was is another question, for the journal was, up to 1830, the product of virtually one individual, and when the (Utilitarian) governorship took it in hand, it went under. There is, however, little doubt as to Brande's own Utilitarianism, no matter how self-serving. He was thoroughly imbued with the ethos of scientific expertise and the professional control of public problems and government matters, and in this sense, the journal was breaking new ground, for it made an obvious contribution to scientific professionalism. It was as specialized as a pre-Victorian periodical could be without going into debt. Furthermore, the journal did find an audience wide enough to make it London's leading scientific publication for many years, and thus Brande was not operating in a vacuum. The Managers did not have any strong interest in changing its content, as subsequent publication revealed, but were instead interested in the journal being more closely identified with themselves. Brande had no intention, prior to their interference, of discontinuing the work, and if the Managers had not mismanaged it financially it could have easily continued to find a viable (and possibly expanding) market for years to come. During the years of publication, however, it served a very important purpose for the development of the RI, reinforcing the image of the Institution as the locus of a certain type of scientific activity; and this was, perhaps, no small achievement.

A third project of Brande's which was clearly Utilitarian, and which continued to identify the RI with the scientific management of social disorder, was the involvement of the Institution in the growing interest in street illumination by gas. That gas illumination spread widely during the Regency was no accident. Certainly, illumination was part of the modernization process, the 'new face' which London was putting on at this time; but it was no less a part of the need to control crime and the seditious forces seen to be threatening the government during these years. Gas illumination meant, for the middle and upper classes, freedom from fear. It was less of an improvement than 'a revolution in public security ...'.[147] One chemical-industrial writer, Colin Mackenzie, noted, as early as 1821, that prior to 1810

> the light afforded by the street lamps hardly enabled a passenger to distinguish a watchman from a thief, or the pavement from the author. The case is now different, for the gas-lamps afford a light little inferior to day-light, and the streets consequently divested of many terrors and disagreeables, formerly born[e] with, because they were inevitable.[148]

The RI itself was very much aware of the role gas illumination would play. Samuel Clegg, a gas engineer whose work was discussed in Brande's journal on

---

[147] G. M. Young, *Victorian England: Portrait of an Age* (2nd ed.; New York: Oxford University Press, 1964), p. 7.

[148] Quoted in Dean Chandler and A. D. Lacey, *The Rise of the Gas Industry in Britain* (London: British Gas Council, 1949), p. 2.

several occasions, pointed out that prior to gas illumination, 'the shades of night offered an easy escape to depredators, by whom the metropolis was infested'.[149] John Millington testified to a Select Committee of the House of Commons in 1823 that the general diffusion of gas lighting would greatly facilitate police purposes.[150] This is not to say that it did not have its fashionable aspect; frequently, as I have suggested, these two views were in easy harmony. Thus Sydney Smith could write to Lady Mary Bennett:

> ... science puts such intense gratification within your reach! Dear lady, spend all your fortune in a gas-apparatus. Better to eat dry bread by the splendour of gas, than to dine on wild beef with wax candles.[151]

But gratification would never have been enough to turn gas lighting into an industry. The phenomenon of gas illumination was discovered in the seventeenth century and widely known in the eighteenth. Lord Dundonald, for example, patented a process of obtaining gas from coal tar distillation in 1781, and lit up Culross Abbey with it six years later. But it was only a converging set of social needs that could render it popular or deem it necessary; and from 1809, owing to such needs, the gas industry emerged in England in full force.

The transformation took place in little more than a decade. In 1814 Benjamin Abbott wrote to Faraday, then on the Continent with Davy, that 'the Gas lights get on here like a house on fire—Westminster Hall & all the streets thereabouts are lighted—Westminster Bridge is contracted for—Bishopsgate Street is also lighted, i.e. the Shops—'.[152] It was a spin-off industry that created numerous jobs in London, and in fact Robert Faraday, Michael's older brother, made his living as an oil gas fitter.[153] But it was not only London that was affected. By 1821, when Mackenzie and Smith were singing the praises of gas, no town in the U.K. with a population of more than 50,000 was without a gas company. Between 1817 and 1830 no less than fifty separate patents were taken out relating to gas manufacture and consumption.[154]

That the Royal Institution became important in the history of gas lighting is to be expected. Although gas manufacture was a purely industrial process—coal tar or oil was destructively distilled and the vapour bottled under pressure—it was one which had various problems which London's leading laboratory might look into. Furthermore, it was connected with police efficiency, a Benthamite project with which the Utilitarian governors would naturally concern

---

[149] Quoted in Samuel Clegg, Jr., *A Practical Treatise on the Manufacture and Distribution of Coal-Gas* (London: John Weale, 1841), p. 5.

[150] Report from the Select Committee on Gas-Light Establishments. Ordered, by the House of Commons, to be Printed, 7 July 1823.

[151] Chandler and Lacey, *Gas Industry*, p. 1. The letter was written in 1821. Smith himself co-patented a gas apparatus in 1844.

[152] *LPW*, I, 83.

[153] *LPW*, p. 107.

[154] M. E. Falkus, 'The British Gas Industry before 1850', *The Economic History Review*, Second Series, XX (1967), 496, and Bennet Woodcroft, *Subject-Matter Index of Patents of Inventions* (London: Eyre and Spottiswoode, 1854), entry under 'Gas Manufacture and Consumption'.

21. The Library of the Royal Institution

22. Early gas lighting: street illumination in Pall Mall, 1807

themselves. And, as Brande was probably one of the first to realize, its importance would inevitably give professional science, in the form of technical
chemistry, a boost, and thus was a problem worthy of Institution interest. In
the course of their work on gas illumination Brande discovered naphthalene in
coal tar (1819), and Faraday benzene (1825).[155]

Brande's first involvement in gas illumination was in 1816, when he supervised the construction of a coal gas apparatus at Apothecaries Hall, which
supplied gas for lighting the building and its laboratories.[156] He also installed a
gas apparatus at the RI the same year, and had John Newman, the Institution's
instrument-maker, design a gas lamp.[157] Yet shortly thereafter, for reasons that
are not clear, Brande declared himself a proponent of oil gas, rather than coal
gas, illumination, and in so doing raised a whole series of issues regarding cost,
efficiency, and pollution. His motives, as some charged, may have been less
than disinterested, and in the end he failed to get the one accepted over the
other; but the major outcome of the (unresolved) argument was to advance the
role of scientific expertise in the technical coordination of social problems.

Initially, Brande was very enthusiastic about coal gas, as his articles in the
journal indicate, and the very first issue reported his own experiments at the
RI, involving a cost-benefit analysis of Newcastle coals as compared with other
substances.[158] His goal was to squeeze maximum energy out of minimum
expense. Thus he carefully discussed Samuel Clegg's 'Horizontal Retort', which
produced, from two guineas' worth of coal, roughly £7.10s. more in profit than
the older retort.[159] In 1818 he had John Millington write an article for the journal, praising the general adoption of coal gas.[160] But it was during the next year
that Brande, without any obvious cause, changed his mind. In 'A few Facts
relating to Gas Illumination', Brande conceded that the

> producing from coal an aëriform fluid, which could be distributed at pleasure in
> every direction, for the purpose of economical illumination, has justly been ranked
> amongst the greatest benefits which the science and enterprise of this country has
> conferred on mankind.[161]

However, he contended, it has significant problems. The high sulphur content
of coal gas causes it to tarnish, and to give off a bad odour, which means it is
useless for home consumption. The apparatus is large and expensive; and coal

---

[155] On naphthalene *see QJS*, VIII (1820), 287–90. On benzene *see LPW*, p. 108, and Faraday,
'On new compounds of carbon and hydrogen, and on certain other products obtained during the
decomposition of oil by heat', *Phil. Trans.*, CXV (1825), 440–66.
[156] Tulley, *Brande*, p. 157. *See also* Brande, 'Some Account of Mr. Samuel Clegg's Improvements of the Apparatus Employed in Gas Illumination', *QJS*, I (1816), 278–83.
[157] *MM*, VI, 79, and *QJS*, VIII (1820), 233.
[158] 'Observations on the Application of Coal Gas to the Purposes of Illumination', *QJS*, I
(1816), 71–9.
[159] 'Further Account of Mr. Samuel Clegg's Improvements of the Apparatus used in Gas
Illumination', *QJS*, II (1817), 132–8.
[160] John Millington, 'On Street Illumination', *QJS*, V (1818), 177–81.
[161] *QJS*, VII (1819), 312.

gas has had an injurious effect on fisheries. Brande pointed out that Messrs John and Philip Taylor had recently started manufacturing gas from oil, and in the next volume of the journal he reported the conversion from coal to oil at Apothecaries Hall. In November of 1819 Brande delivered the Bakerian Lecture to the Royal Society reporting his findings on the relative merits of coal and oil, claiming that oil gas gave nearly three times the illumination, and consumed only one half as much oxygen, as did coal gas. Much of the laboratory work, he added, was done by 'Mr. Faraday, whose accuracy as an operator is not inferior to his assiduity as my assistant in the Laboratory of the Royal Institution ...'.[162] In general, Faraday and Brande demonstrated that the plants of most commercial gas companies were 60% efficient.[163] Faraday wrote to Gaspard de la Rive 'that the gas is a purer and more highly carbonated substance and that the whole operation is performed in a clean and neat manner and with very little trouble'.[164] By 1820 a number of court trials had arisen over the subject, and by 1824 there was a general attempt to replace coal with oil in Westminster illumination.[165]

In general, oil gas came to be identified with the Royal Institution. As already noted, Faraday's brother was an oil gas fitter, and the RI installed oil gas in 1828.[166] David Gordon, who patented the portable gas lamp, had Faraday do the analysis of a fluid that collected in the receivers—marking the discovery of benzene. J. F. Daniell patented a process for manufacturing gas from resin in 1827, which Gordon's Portable Gas Company subsequently adopted. Benjamin Hawes, a noted Utilitarian and Member of the RI, installed an oil gas apparatus in his soap-works.[167] Testimony given in 1824 in favour of the London and Westminster Oil-Gas Co. consisted entirely of RI personnel, viz. Brande, Faraday, and Richard Phillips.[168] Finally, Brande managed to raise enough of a furor over the polluting properties of coal gas so as to induce RI governor William Sturges Bourne, then Home Secretary, to appoint Brande, Roget, and Thomas Telford commissioners to investigate the problem of water supply, in which John Bostock assisted.[169] 'There can be no doubt,' Brande stormed in a journal editorial,

[162] 'On the composition and analysis of the inflammable gaseous compounds resulting from the destructive distillation of coal and oil, with some remarks on their relative heating and illuminating powers', *Phil. Trans.*, CX (1820), 11–28; the quotation is on p. 17.

[163] Tulley, *Brande*, pp. 150–1.

[164] *LPW*, I, 111–12; dated 6 October 1818.

[165] ibid., p. 115. On the London and Westminster Oil-Gas Co. *see* William Matthews, *An Historical Sketch of the Origin, Progress, & Present State of Gas-Lighting* (London: Rowland Hunter, 1827), pp. 166–227.

[166] Chandler and Lacey, *Gas Industry*, p. 71 and note.

[167] Matthews, *Historical Sketch*, p. 185. Hawes (Sir Benjamin as from 1856) was active in the Diffusion Society and the Political Economy Club, and as I. K. Brunel's brother-in-law supported the Thames Tunnel project and became very interested in the establishment of railways and telegraphs.

[168] ibid., pp. 169ff. Faraday, however, expressed some reservations.

[169] *See* their letter in the *London Medical Gazette*, II (1828), 271–5 and 305–8, which attacks coal gas manufactories as a major source of pollution; and John Rickman (ed.), *Life of Thomas Telford, civil engineer, written by himself* (London: James and Luke G. Hansard and Sons,

that an atmosphere tainted by coal gas is injurious to animal and vegetable life ...
To say nothing of danger from fire and from explosion, it has always been [a] matter
of surprise to us that gas-works are tolerated by the government in close and confined situations—that the Thames is suffered to be polluted with their offal, and that
they are sometimes place[d] close by the road side ... to the nuisance of every one
who passes. These matters want looking into.[170]

All this was not without opposition from the coal gas companies, and in the
end they did manage to prevail, though it was never made clear whose assessment of the dangers and efficiency of the two gases was correct. George Lowe, a
leading coal gas engineer, attacked Brande's 'few Facts relating to Gas
Illumination' as 'a most extraordinary fact-perverting paper', and pointed out
that the article ignored progress that had been made in gas purification.[171]
William Matthews charged that Brande had a vested interest in oil gas, and
that he gave false testimony to the Select Committee in 1824 appointed to consider the replacement of coal gas by oil gas.[172] William Herapath reported that
experiments on the two gases revealed no difference in odour.[173]

For a while, it seemed as though oil gas would win the day. In 1822, for
example, Sir William Congreve, under appointment by the Home Secretary,
reported on the state of London gas works. After stressing the importance of
the subject for police purposes, Congreve stated that oil gas was less explosive,
brighter-burning, better-smelling, and less polluting than coal gas; and,
moreover, that it was cheaper.[174] In 1824, the Postmaster General of Ireland
reported the replacement of coal gas by oil gas. The former, he stated, 'was
attended with considerable annoyance to the officers of this establishment,
from its offensive effluvia and suffocating quality'.[175] But oil gas did not replace
its predecessor, and we can only guess that by the mid-1820s the coal gas companies were too powerful to be defeated. The London and Westminster Oil-
Gas Co. lost its bid for incorporation, and coal gas continued to be the major
source of illumination. By 1834, the RI itself switched back to coal gas lighting.

In retrospect the debate seems to have been of little moment, for the two
sides had far more in common than apart. The argument was a material conflict rather than an ideological one, for both sides were operating from the

1838), esp. pp. 627–32. For the official documents *see* Copy of the Commission issued for in-
quiring into the State of the Supply of Water to the Metropolis, printed 17 March 1828; Report
of the Commissioners Appointed by His Majesty to inquire into the State of the Supply of Water
in The Metropolis, 21 April 1828; and Report of the Select Committee on the Supply of Water to
The Metropolis, 19 July 1828.

[170] 'To Our Readers and Correspondents', *QJS*, XXIV (1827), unnumbered page following
title page.

[171] George Lowe, 'Remarks on an Article entitled "A few Facts relating to Gas Illumination,"
published in No. 14 of The Quarterly Journal', *Phil. Mag.*, IV (1820), 37–46.

[172] Matthews, *Historical Sketch*, pp. 111–12, 176–7, and 186.

[173] William Herapath, 'Experiments on Oil and Coal Gas', *Phil. Mag.*, LXI (1823), 424–31.

[174] Report on the State of the Gas Works in the Metropolis. Ordered, by the House of Commons, to be Printed, January 1822. Further Report, January 1823.

[175] GPO. Returns relating to the Lighting with Oil Gas the General Post Offices of Great Britain and Ireland. Ordered, by the House of Commons, to be printed, 5 May 1824.

same philosophical base. Efficiency, expertise, the translation of social problems into technical ones—all this was common to them both. The image of science as the ground of social coordination was much advanced by the debate, and the RI was seen as the natural centre of a controversy involving cost-benefit analysis, technical consultancy, litigation and patents, government testimony, and problems of pollution. If the Institution had picked the 'wrong side' of the debate, its role in the advancement of street illumination, and more generally social-technical innovation, was nevertheless enhanced. Even the role of scientific theory in such matters, and not just scientific method, was reinforced by the whole affair. Samuel Clegg's son, also an engineer, devoted an entire chapter of his *Practical Treatise* (*see* above, footnote 149) to 'Chemistry, as Applied to the Manufacture of Coal Gas', in which he urged the gas engineer to familiarize himself with caloric theory. Of course, what he was doing was giving a description of gas illumination in terms of contemporary theory, i.e. after the fact. Knowledge of caloric theory was hardly a contributing factor to technical innovation in metropolitan street lighting. Yet Clegg was not alone in believing that it was, and the pre-Victorian confusion over definitions of science helped entrench that belief, as well as boost its organization and professionalization.

The RI also lost the battle for water purity, of which the coal gas question was only a part. At the same time that Brande was attacking the companies for polluting the Thames, a man named John Wright published a muckraking pamphlet *The Dolphin* (1827), which revealed that the intake ('dolphin' in the trade) of the Grand Junction Company, a leading water supplier, was located a short distance from the outpouring of a major sewer. When Brande, Roget, and Telford published their report the following year, it confirmed Wright's charges and added most of the other water companies to the list of culprits. It recommended that their monopoly be broken, and that their operations 'be subjected to some effective superintendence and control'.[176] Despite the urgent demands of many voices, principally of medical professionals, for speedy action, the government took no action at all, and the issue was dropped until the late 1840s. Nevertheless, the loss of this campaign again had important positive aspects in terms of the public image of science and the Royal Institution. The political agitation generated by Wright's pamphlet, and the Brande-Roget-Telford report, really mark the beginning of the sanitary reform movement. There was extensive editorial comment in a large variety of papers and journals, and all the medical publications discussed it.[177] Brande delivered a Friday Evening Discourse on the subject on 23 January 1829.[178] Faraday, who had been doing water analyses since 1815, soon found himself a recognized expert

[176] 'Supply of Water in the Metropolis', *The Lancet*, 21 June 1828, p. 377. The story of *The Dolphin* and the subsequent inquiry is discussed in G. Phillips Bevan, *The London Water Supply* (London: Edward Stanford, 1884), pp. 14–20, and Daniel E. Lipschutz, 'The Water Question in London, 1827–1831', *Bulletin of the History of Medicine*, XLII (1968), 510–26.

[177] Lipschutz, 'Water Question', pp. 520 and 524.

[178] *QJS*, n.s., V (1829), 350–6.

on the subject of water purity. Parliament sought his testimony on sewers in 1834, and Chadwick chose him to do the analyses for the Health of Towns Commission in 1843. When Faraday wrote his famous letter on the pollution of the Thames to *The Times* in 1855 (*see* Plate 27), he was in fact following the pattern of a life-long career that had been shaped by these events, and by the drive for medical professionalization.[179]

Before turning to Faraday's role in the evolution of the Royal Institution, it might be helpful to point out the profound differences between him and Brande as regards the political aspects of their work. Despite the fact that Faraday found himself heavily involved in Brande's Utilitarian causes, he was largely unaware of his mentor's purposes. Faraday did not regard such work as public service (and he was a loyal subject) or Institutional obligation (and he was a faithful employee) as a means of creating a professional role for science. By the time he inherited the mantle of metropolitan scientist, he had become a Utilitarian *malgré lui*. Yet such personal lack of interest made no difference for the actual course of the social uses of science.

The contrast between Brande and Faraday in these matters is perhaps most clearly revealed in their various testimonies before Select Committees and Royal Commissions. These were often concerned with matters of social engineering, and reveal the mentality of the commissioners vis-à-vis science. How Brande and Faraday each responded to that mentality is remarkable. Brande knew what his questioners wanted to hear and was eager to please, even to the point of giving false testimony (perhaps 'exaggerated' might be more euphemistic). On a number of occasions this created embarrassing situations, for Brande was trying to create a role for scientific expertise whether it was relevant to the particular problem or not. In 1816, for example, Lord Althorp, son of the Earl Spencer, then RI President, raised a question in the House of Commons regarding restrictions on leather manufacture.[180] Since the matter had important repercussions in the industry, a Select Committee was appointed, and interviewed Brande on June 3rd, an interview that probably took up most of the day.[181]

The issue was as follows. It was the tanner's job to convert hides into rough

---

[179] For examples of Faraday's water analysis work *see* RI Archives, Laboratory Notebook for 1813–21, p. 27, and the notebooks for 1821–9, 1830–59, *passim*. An interesting report on various waters is in Letter Box FL-1, dated 5 April 1837. Williams discusses this work in *LPW*, p. 106. On the 1843 appointment *see* S. E. Finer, *The Life and Times of Sir Edwin Chadwick* (London: Methuen, 1952), p. 233. Faraday's letter to *The Times* was printed on 9 July 1855.

[180] Althorp became an active Utilitarian. He was elected to the Political Economy Club and was 'influential in opening the door to the Benthamites'. Both Charles Shaw Lefevre (an RI governor and well-known Utilitarian) and Edwin Chadwick were drawn into government service through him (*see* S. E. Finer's essay in Gillian Sutherland, *Studies in the Growth of Nineteenth Century Government*, p. 17). Perhaps equally important, the Spencer estates, as mentioned in Chapter 2, were located in Northampton, where tanning and shoe manufacture were vital industries.

[181] Report from the Select Committee on the Leather Trade. 3rd Report. Printed by order of the House of Commons, 6 June 1816. In what follows I have drawn on this and the article by C. H. Spiers cited in footnote 102.

tanned leather, or 'crust' leather, and the currier's to convert this into leathers for specific uses by cleansing, shaving, oiling, etc. Tanners now wanted the right to by-pass this dependency on curriers by being allowed to cut the hides into sections and to tan different pieces in various ways, appropriate to different uses. Although it was logical to call on Brande to testify, inasmuch as he was London's representative of commercial chemistry, he did not have any real familiarity with leather manufacture. Unlike Davy, Brande had not worked directly with tanners, and there is no evidence that he had any specialized knowledge of leather chemistry. Yet what emerges from the testimony is his haste to propagandize for the indispensability of scientific expertise. 'Have you,' he was asked, 'devoted much time and attention to gain a theoretical and perfect knowledge of the art of tanning, with a view to improve the manufacture of leather?' Brande replied in the affirmative, then proceeded to reveal a knowledge that was very scanty. He took the side of the tanners, arguing that if they tried to tan different parts differently under the present regulations, the curriers would object. The testimony continues:

> Why is that objection made; from prejudice, judgment, or what?—I should presume entirely prejudice.
> Do you know anything of currying?—No, I do not.
> Why should you suppose the currier would be more prejudiced in his judgment than the tanner?—I have seen very good leather tanned with different materials than those generally in use, which I was told the currier refused to purchase.
> Are you sure it was good leather?—Yes.
> How did you know it was good leather?
> —It had all the characters of perfectly tanned leather.
> Do not you know that the proof of whether dressing leather is good or bad, is in the currying?

Brande had painted himself into a corner: obviously if the proof of quality was in the currying, it was foolish to claim that the crust leather could be judged at a glance. He could only respond, quite feebly, 'I cannot give an opinion upon that subject'. Soon after he claimed that chemical experiments were sufficient to test the quality of leather, but this too was an example of wrong-headed professional zeal: there was no way, as Dr Spiers points out, that the chemistry of 1816 could tell how well or poorly a a sample of leather would wear over time.

That chemistry was largely irrelevant to quality control emerged as an explicit point in the testimony, once again to Brande's embarrassment. After he stated that English leather was the world's best, Brande was asked a crucial question:

> Have not the French chemists a very good knowledge of the tanning process?—One or two of the French chemists have certainly thrown a good deal of light upon the process.
> Then how do you account for the fact, that, without any restrictions in France, the French leather has not excelled the English?—That is a question which I could only answer after having seen a French tan-yard.

Precisely: the problem of quality was a matter of craft, not chemical theory. By the end of the interview, Brande had completely undercut not merely his own claims to expertise, but revealed that these types of issues could not be resolved by the observations of the laboratory. In spite of this, scientific testimony (Brande's included) became an increasingly popular part of the legislative process. The role which science was called upon to play, and its importance for the Utilitarian vision, carried the day over the reality of its limitations.

Brande's false confidence stands in sharp contrast to Faraday's diffidence in such matters. On 11 July 1834, for example, he had to testify to a Select Committee on sewers. The Committee included a former RI governor, Henry Warburton, and Benjamin Hawes (mentioned above), both strongly Utilitarian in their politics. Faraday was consistently naive in the face of their leading questions, giving what they undoubtedly considered the wrong answers. In general, his testimony gave the impression that he did not understand what he was doing there. He emphasized that his knowledge was only theoretical, and that he simply did not know enough about sewers to provide any useful information.[182]

The following year, both Faraday and Brande were called upon to testify on the problem of ventilating the Houses of Parliament. Again, two things emerge as remarkable: the disparity of their testimony, and the mentality of their questioners. Benjamin Hawes was in the chair, and Warburton again on the Committee. The interviewers wished to 'arrange' the air, and their line of questioning reveals a strong commitment to the notion of the technological fix. They specifically wanted to use chemistry to restore a depleted atmosphere to its original state, and Faraday, at least, repeatedly pointed out not merely his own ignorance of ventilation but the limited ability of chemistry to carry out such a task. He argued that the ventilation of the RI's own Lecture Theatre was not very successful, that his own plan produced partial drafts, and that ascertaining 'the chemical effect produced upon a body of air in a chamber crowded for many hours' depended 'on circumstances beyond the present reach of chemistry ...'.[183] The attempt to restore a vitiated atmosphere, he told Hawes, was not within the scope of contemporary chemical knowledge. Time and again, Faraday declined to reply in the affirmative to leading questions that encouraged him to say that science was relevant to these matters. When the Committee finally asked, 'would it not be desirable to combine with the architect some scientific man in order to obtain the best results?', Faraday indicated that that was a choice which the Committee members would have to make for themselves.

This testimony, given on 7 August 1835, was followed by that of Brande five days later which constituted almost a complete reversal. Brande understood his

---

[182] Report from the Select Committee on Metropolis Sewers. Ordered, by the House of Commons, to be Printed, 8 August 1834.
[183] Report from the Select Committee on the Ventilation of the Houses of Parliament. Ordered, by the House of Commons, to be Printed, 2 September 1835.

examiners well and welcomed an opportunity to promote the role of science in public affairs. His tone was bold and self-assured. He gave exact answers and plans, and tossed out atmospheric temperatures and ventilation velocities with ease. He also gave testimony that was patently false, and as in the leather testimony of nearly twenty years before managed to get caught. Thus he suggested ventilation via a few large apertures and denied the possibility of producing drafts. Later, he was forced to admit that several RI Members had complained of drafts in the Lecture Theatre. He also claimed that air purity could be tested chemically, and stated that he had been asked to make such a test years ago at the Covent Garden Theatre. Despite the occasional slip, it was a good bit of salesmanship, precisely the type of performance his examiners desired and expected. And it was this sort of expectation that carried the day. As we shall soon see, it did not matter what a scientist like Faraday personally believed science was, or should be, about.

To sum up, in the two decades after the electoral changes of 1811, the Royal Institution began to establish for itself the public role of yet another scientific ideology. Of necessity, it had to move against the background of the amateur tradition, which retained its greatest influence in the areas of lecture programmes and financial arrangements; but the RI's shift towards expertise and professionalism was of immense historical significance. Thanks to Brande, the Utilitarians, and the activities they fostered—textbooks, a journal, consultancy, gas illumination, commercial analyses, medical qualification, legal and government testimony—the RI began to take on the image of a metropolitan powerhouse for the scientific management of social problems. It was within the context of this momentum that Michael Faraday took charge of the Institution; and it is only within such a context, I believe, that the life and work of one of Britain's greatest scientists can properly be evaluated.

5
A Portrait of Michael Faraday

Temperamentally, Michael Faraday was the least likely member of the British scientific establishment to inherit the mantle of metropolitan scientist. His comments on professional and commercial work were unequivocally negative, and his orientation to life, as several biographers have noted, was profoundly religious and other-worldly. Yet he became, as Pearce Williams has remarked, 'a trouble-shooter of the first rank' in public and technical affairs.[1] We have already seen how Faraday was Brande's faithful assistant, laboratory analyst, partner in the medical lectures, gas illumination work, and (for a time) the journal; and how the Friday Evening Discourses served as a 'message centre' regarding recent developments in modernization. What I hope to do in this discussion is not to discover a new Faraday, but elaborate on the nature and the conflict of the 'inner' and the 'outer' man. How mystical Faraday was has perhaps escaped sufficient attention; how extensive his work was in social engineering and technical-industrial analyses seems equally ignored. That these two Faradays, so to speak, were antagonistic to each other is at least one point I wish to make. In Brande's case, one might conceivably (but I think incorrectly) argue that his professionalism or Utilitarianism was a personal affair, not part of the governors' collective mentality or the Institution's emerging social role. But Faraday is the acid test, for neither his religious mysticism nor his aversion to 'trade', as he once called it, prevented him from becoming London's leading technological 'fixer', certainly outstripping Brande in terms of work load. It is precisely the anomalous nature of Faraday's career, the acuteness of the dichotomy between his personality and his social function, that casts the nature of the Royal Institution into sharp relief. In drawing a portrait of Faraday, then, I hope once again to demonstrate, as was the case with Davy, that the governors, and not the scientists, were the 'real' Royal Institution.

The single most important fact about the 'inner' Faraday was a deep religious commitment that pervaded his life and work, and which was derived from Sandemanianism, the religious sect founded by John Glas (1695–1773). Auguste de la Rive, commenting on the agitation that upsets the lives of most men, wrote to Faraday in 1852: 'Tout cela m'amène à penser au malheur de

[1] *LPW*, p. 483.

ceux qui n'ont pas cette foi religieuse que vous avez à un si haut degré'.[2] In his éloge of Faraday, the first biographical sketch published, de la Rive emphasized that 'Faraday was, in fact, thoroughly religious, and it would be a very imperfect sketch of his life which did not insist upon this peculiar feature which characterized him'.[3] Faraday was completely persuaded by John Glas' favourite quotation from the Bible: 'Thy Kingdom is not of this world'.[4] Thus Williams points out that Faraday's 'deepest intuitions about the physical world sprang from this religious faith in the Divine origin of nature',[5] and Silvanus Thompson correctly stated that 'no review of his life-work would be complete without a fuller reference to the religious side of his character'.[6] Faraday himself denied any relationship between his science and his religion, but an analysis of the two does not bear out this denial.

It was Glas, a defrocked Scottish minister who formed his own church at Dundee, whose works constituted the intellectual basis of the Sandemanian sect.[7] Glas was later at Perth, where he was joined by Robert Sandeman (1718–71), a convert and future son-in-law. The proliferation of fundamentalist sects in the eighteenth and nineteenth centuries, accelerated by industrialization and social dislocation, is a well-known phenomenon, and Glas was led to assert that a literal reading of the Bible was the only sure guide to life. Yet in an age of increasing materialism, Glassism represented a phenomenon almost directly opposite to that which Weber described in *The Protestant Ethic and the Spirit of Capitalism*. Wealth and worldly success, if Christ was to be taken at his word, were not evidence of membership in a predestined elect, but largely a waste of time. This also meant that our purpose on earth was a divine one, other-worldly. Faraday's contempt for making money, for the world of trade, and for any form of democratic or even mildly progressive politics, derives from his early socialization by the Sandemanian church.

Christ's dictum, 'Lay not your possessions up where moth and rust decay', was deeply felt by the Sandemanians. Of what, then, did a man's wealth consist? Clearly, in his faith, his relationship with God. Yet again, Glas was most atypical on this point. He had no interest in the sort of revival fervour characteristic of Methodism, for example. He was, instead, rather dry and matter-of-fact, defining faith as 'a bare intellectual acceptance of certain facts'.[8] This left, however, an ambiguous legacy. Faith was an intellectual activity, and conversely intellectual activity was a religious activity; but for this very reason

---

[2] *BJ*, II, 317.

[3] Translated as 'Michael Faraday, his Life and Works', *Phil. Mag.*, XXXIV (1867), 409–37; the quotation is on p. 411.

[4] J. G. Crowther, *British Scientists of the Nineteenth Century* (London: Routledge and Kegan Paul, 1935), pp. 69–126, *passim*. The religious background to Faraday is discussed in James F. Riley, *The Hammer and the Anvil* (Clapham, via Lancaster, Yorks.: The Dalesman Publishing Company, 1954). *See also* H. Marryat, 'A Student of Scripture', *The Times*, 21 September 1931, p. vii.

[5] *LPW*, p. 4, and *see also* pp. 103–6.

[6] S. P. Thompson, *Michael Faraday* (London: Cassell, 1898), p. 5.

[7] On Glas see Alexander Gordon, 'John Glas', *DNB*, XXI (1890), 417–18.

[8] ibid., p. 417.

science and religion did not need each other for their substantiation. Thus Glas rejected John Hutchinson's claim to have discovered a system of physics in Holy Scripture, stating that 'the Bible was never designed to teach mankind philosophy'.[9] Faraday echoed this in a famous letter to Lady Lovelace in 1844, which has been quoted on numerous occasions;[10] but Faraday's statement that science and religion should be kept separate was not a rejection of their central unity—in which he believed—but of the argument by design, which the Sandemanians regarded as superfluous.[11] In this, once again, they were atypical, for the English scientific tradition, from the seventeenth century onwards, had been 'Through Nature up to Nature's God', and was kept strongly alive in the natural theology tradition of the first half of the nineteenth century.[12] Faraday's faith was so total, so unquestioning, that the idea of proving what was obvious seemed to him absurd. As Williams has pointed out, for Faraday the pursuit of science was not a proof of God, but a participation in His Divinity.[13]

Glas' belief that faith was the 'bare intellectual acceptance of certain facts' was, I believe, the cornerstone of Faraday's inner reality. These facts, for Faraday, consisted (among other things) of a scientific principle that was becoming evident from a variety of sources, including German *Naturphilosophie*, work equivalents demanded by the Industrial Revolution, and internal developments within science itself. Yet it is hard to state whether the 'correlation of forces' idea was, in the early nineteenth century, scientific in a truly empirical sense. All the forces of nature, Faraday believed, are identical[14] (and therefore interconvertible) and derivative from the one force of all Nature, viz. God. Faraday's intellectual achievements revolved around the essential identity of forces: the motor, the generator, the identity of electricity experiments, the chemical equivalent of electricity, electromagnetic induction, specific inductive capacity, the rotation of the plane of polarized light in a magnetic field, para- and diamagnetism, the search for an identity between gravity and electricity, and his last experiment, in 1862, the search for the splitting of spectral lines in a strong magnetic field (later discovered by Zeeman). Not content with explanations in terms of general fields of force, Faraday began, in 1844, to argue that matter itself was an arrangement of forces, hence his flirtation with Boscovichean atomism. The monolithic vision which Faraday possessed makes him, in fact, the speculative scientist par excellence; and it accounts for his ability to create a major scientific synthesis even though constantly being distracted—which was, as we saw, precisely where Davy failed. Despite Faraday's constant inveighing against hypotheses, he himself had his thesis before he began to consult the data and never wavered to the end

[9] ibid.

[10] *BJ*, II, 195–6.

[11] Sandeman, however, did use the argument in his book, *Some Thoughts on Christianity* (1764).

[12] C. C. Gillispie; *Genesis and Geology* (New York: Harper and Row, 1959).

[13] *LPW*, p. 104.

[14] *BJ*, II, 484.

of his life. Between 1821 and 1831, when there was no evidence whatsoever that one electric current could induce another, he did experiment after experiment trying to prove this. He would carry magnets and a coil in his pocket, and in his spare time play with them and ponder over what arrangement would demonstrate the relationship that he was convinced existed. The experiments with polarized light were pursued with the same sense of religious dedication, as was the (unsuccessful) attempt to establish a relationship between gravity and electricity.[15] Although there was a quiet intensity to all his work, there was no fanatic zeal, for to Faraday there was no problem: the thesis simply could not be wrong. At the end of a series of negative results, Faraday wrote that they 'do not shake my strong feeling of the existence of a relation between gravity and electricity . . .'.[16]

Faraday's mysticism, if so it can be called, was thus not a belief in occult powers, and in fact he regarded the numerous letters he received from mediums and spiritualists as unmitigated rubbish.[17] But he did believe in the existence of a spiritual entity as the source of all manifest physical forces, and saw every instance of interconvertibility as just one more case of what was so obviously true. Although, with the exception of one document (*see* below), Faraday did not develop this into any type of sustained argument, he stated this position on a number of occasions.

Faraday's earliest lectures, e.g. to the City Philosophical Society, contained the thesis that the forces of nature were one and the same,[18] and this was repeated at many points. At the RI in 1834 Faraday said of chemical affinity, electricity, heat and other forces: 'We cannot say that any one is the cause of the others, but only that all are connected and due to one common cause'.[19] The sketch of Faraday in *Men of the Time*, presumably written by Faraday himself in 1852 (or possibly based on an interview), stated that it 'is the hope of this philosopher that, should life and health be spared, he will be able further to aid in showing that the imponderable agencies just mentioned are so many manifestations of one and the same force'.[20] Occasionally, Faraday related this hypothesis to God. Thus at the Friday Evening Discourse of 10 June 1836, he stated that there may be 'some *more high and general power* of nature even than electricity . . .'.[21] When Albert visited the RI on 26 February 1849 for a private lecture, he heard Faraday speak in virtually eschatological terms regarding the ultimate meaning of magnetism. 'What its great purpose is,' he told the Prince Consort,

[15] Of gravity Faraday wrote in 1849: 'Surely this force must be capable of an experimental relation to electricity, magnetism, and the other forces, so as to bind them up in reciprocal action and equivalent effect. Consider for a moment how to set about touching this matter by facts and trial' (*BJ*, II, 252).

[16] *BJ*, II, 253.

[17] A number of these are preserved in the Faraday Papers at the IEE. *See* also *BJ*, II, 307–8, 318, and 468–70.

[18] *BJ*, I, 212–13.

[19] ibid., II, 47.

[20] *Men of the Time* (London: David Bogue, 1856), p. 272. Faraday did the proofs of this, and his MS notes are in the RI Archives, Letter Box FL-4.

[21] *BJ*, II, 86; italics in original.

seems to be looming in the distance before us, the clouds which obscure our mental sight are daily thinning, and I cannot doubt that a glorious discovery in natural knowledge, and of the wisdom and power of God in creation, is awaiting our age . . ..[22]

Undoubtedly, Albert regarded these remarks as 'good form', typical expressions of Victorian piety. But Faraday almost never went in for 'form'.

The link between the forces of nature and God emerges most strongly in Faraday's speculations on the nature of matter itself, which he first made public in 1844. At this time, he was just emerging from the effects of a nervous breakdown, which had forced him, from 1839, to withdraw increasingly from Institution activities. He stopped giving Friday Evening Discourses and carrying on other RI activities in 1841, stopped having visitors in 1842, and did no scientific work for all of 1842 and 1843. What happened to his mind during this time is of course unclear, but that he emerged in 1844 into a new burst of creative activity is a matter of historical record. And, probably not coincidentally, his scientific research and his religious convictions had, by 1844, meshed completely.

Faraday returned to the Discourses on 19 January 1844, the first discourse of the season, with a lecture that, for the first time, included a statement on Boscovichean atomism, arguing that we never know an atom apart from its forces.[23] Later that year, in an impromptu talk, Faraday gave the first hint of the electromagnetic theory of light. Radiant phenomena do not need an ether, he stated; instead, the vibrations of heat and light consist of vibrations along the lines of force.[24] Finally, in February of 1845 Faraday sat down and made some notes to himself in a document discovered only a few years ago, discussing his theory of matter and reasons for his belief.[25] Matter is finally eliminated from Faraday's physics; his atom has become completely spiritual, and in support of the force-centred atom he cites God three times. 'Is the lingering notion which remains in the minds of some,' he asks at the end, 'really a thought, that God could not just as easily by his word speak power into existence around centres, as he could first create nuclei & then clothe them with power?'[26] The world has at last been reduced to a comprehensible unity: the Boscovichean atom is a one-force entity that orders all phenomena at the ultimate level. Faraday's *Urkraft* is not derivative from *Naturphilosophie*, but from the Sandemanian (and his own) conception of the Deity. It is not

[22] ibid., p. 244.

[23] 'A Speculation Touching Electric Conduction and the Nature of Matter', discussed in John Tyndall, *Faraday as a Discoverer* (New York: Thomas Y. Crowell, 1961 reprint), pp. 144ff. *See also BJ*, II, 178. The full statement of Boscovich's view of matter is contained in his *Theory of Natural Philosophy* (Cambridge, Mass.: The M.I.T. Press, 1966, from the text of the Venetian edition of 1763). Williams discusses Faraday's Boscovicheanism at various points in *LPW*, but his thesis is highly controversial.

[24] 'Thoughts on Ray-vibrations', *Phil. Mag.*, XXVIII (1846), 345–50.

[25] Trevor H. Levere, 'Faraday, Matter, and Natural Theology—Reflections on an Unpublished Manuscript', *British Journal for the History of Science*, IV (1968), 95–107.

[26] ibid., p. 107.

necessary to argue, as Levere points out, that religion shaped the content of Faraday's physics (though that is likely enough), but only that it played a role in his selection of this atomic hypothesis. The 'main reason for his adherence to the hypothesis,' says Levere, 'was that it fitted in with the world picture imposed by his religion'.[27] With matter no longer a problem, the age-old dichotomy between mind and body, matter and spirit, collapses. Microphenomena, *a fortiori* macrophenomena, are various (interconvertible) manifestations of His presence in the world.

It was his insistence on the reality of force fields that made Faraday incomprehensible to the scientific world, which was largely preoccupied with squeezing the forces of nature into the Newtonian, or at least into a mechanical, paradigm. Maxwell has usually been credited with making field theory acceptable by mathematizing it, a process that entailed the substitution of mechanical models for lines of force and claiming only heuristic value for them. In the context of Victorian science, this was quite necessary, for Faraday's mode of understanding had become scientifically unacceptable, being as it was a personal intuition of the nature of ultimate reality. When one of the oldest Members of the RI, Dr Thomas Mayo, pointed out the Berkeleyian logic Faraday was using, he hit the nail on the head. 'Your atmosphere of force,' he wrote to Faraday in 1844,

> grouped round a mathematical point, is not, as other hypothetical expressions have been in the course of your researches, an expression linking together admitted phenomena, but rather superseding the material phenomena which it pretends to explain.[28]

In short, the hypothesis did not allow either verification or falsification, which did not constitute a problem for Faraday in that the single force field present in each and every atom—i.e. God—was not hypothetical. The *Urkraft* was a given, a fact, and thus an obvious explanation of the material world. In 1847, at the last lecture in a course Faraday gave at the RI, he stated this in the most unequivocal manner possible, in a way that was (*pace* John Glas) lyrical, even drunk with fervour:

> Our philosophy, feeble as it is, gives us to see in every particle of matter, a *centre* of force reaching to an infinite distance, binding worlds and suns together, and unchangeable in its permanency. Around this same particle we see grouped the powers of all the various phenomena of nature: the heat, the cold, the wind, the storm, the awful conflagration, the vivid lightning flash, the stability of the rock and the mountain, the grand mobility of the ocean, with its mighty tidal wave sweeping round the globe in its diurnal journey, the dancing of the stream and the torrent; the glorious cloud, the soft dew, the rain dropping fatness, the harmonious working of all these forces in nature, until at last the molecule rises up in accordance with the mighty purpose ordained for it, and plays its part in the gift of *life itself*. And therefore our philosophy, whilst it shows us these things, should lead us to think of Him who hath

---

[27] ibid., p. 101.
[28] Quoted in *BJ*, II, 180.

wrought them; for it is said by an authority far above even that which these works present, that 'the invisible things of Him from the creation of the world are clearly seen, being understood by the things that are made, even His eternal power and Godhead'.[29]

In 1844, the search for a unifield field had been resumed once more and continued to the end of his life. Faraday was, quite literally, at play in the fields of the Lord.

Given the sublime nature of Faraday's interest in science, we can only wonder what he thought of his social role when, for example, he sat before the Select Committee on Metropolis Sewers. As already stated, Faraday was never ambiguous about his attitude to such work. As a youth he spoke of 'Trade which I hated, and science which I loved ... '. His professional duties were 'extraneous matters', and he once told a publisher that 'because my occupation is entirely personal I cannot afford to get rich'.[30] In the 1850s, when Faraday was engaged in a great deal of consultancy work, he replied to a request to analyse some bark in a way that reflected wishful thinking rather than contemporary reality. 'I am not Professional,' he told his correspondent, 'and do not undertake analyses for any one—not even for the Government . . ... There are plenty of professional chemists . . ..'[31] He himself stated that 'I have always felt my position in the Institution as a very strange one'.[32] Strange or not, the volume of commercial-professional work done by Faraday during his RI association (i.e. most of his life) was staggering. It began in 1813, when Brande set him the problem of extracting sugar from beetroot.[33] Early in 1818 Faraday wrote to Benjamin Abbott, 'I have been little any where except on business . . . I have been more than enough employed. We have been obliged even to put aside lectures at the Institution . . .'.[34] By 1820 he was so inundated that he wrote to Sarah Barnard, his future bride, 'as I ponder and think on you, chlorides, oil, Davy, steel, miscellanea, mercury, and fifty other professional fancies swim before me . . .'.[35] The same frustration emerged in a letter to André-Marie Ampère a few years later, in which Faraday described himself as 'an anchorite in the Scientific world', 'all day long in my Laboratory', not however, engaged in original research, but 'unfortunately occupied in very commonplace employment . . .'.[36] 'At one period,' wrote Margery Reid of the 1820s, 'he had to make so many commercial analyses of nitre for Mr. Brande, after preparing his lecture, that it was very late before he could return to his own researches.'[37] In 1830 Faraday wrote to Mitscherlich:

[29] ibid., pp. 229–30.
[30] ibid., p. 365, and Thompson, *Michael Faraday*, p. 234.
[31] Letter to A. Young, 3 January 1855, Wellcome Institute of the History of Medicine, File 69293, Item 67430. Cited by courtesy of the Wellcome Trustees.
[32] Quoted in J. H. Gladstone, *Michael Faraday* (London: Macmillan, 1872), p. 107.
[33] Thompson, *Michael Faraday*, p. 15; *BJ*, I, 57.
[34] *LPW*, I, 109.
[35] *BJ*, I, 322; *see also* p. 284.
[36] *LPW*, I, 153–4.
[37] *BJ*, I, 425; cf. Gladstone, *Michael Faraday*, p. 28.

My situation is such that I am liable to be constantly disturbed and for 8 or 9 months in the year am thoroughly tired with continued business Although I date from the R[oyal] Institution I am at this moment a little way from town resting from recent fatigue & recovering from ill health.[38]

The 1830s may have been slightly less taxing, yet Faraday did not abandon professional testimony in the courts until 1834, Excise work until after 1835, and the medical lectures until 1839.[39] He was even doing technical work during the period of his nervous breakdown. Both he and Brande were on a heavy schedule of professional consultancy, government service, Parliamentary testimony, and the like, throughout the 1840s. In 1844, when two pharmaceutical chemists proposed a school of practical chemistry at the RI, they made a point—since they did not wish the school to be run by Brande or Faraday—of stating that its professors may not be allowed 'to appear professionally in courts of law, or to enter into any such professional occupation, or give any professional advice or opinion, except to the Government of the country'.[40] In the margin of the manuscript draft of the proposal John Barlow, then RI Secretary, added a note also excluding 'attending trials to give evidence for the Excise &c'. Brande and Faraday were, at the time, doing tests on the chemical contents of spirits and liquors for the Admiralty, but Barlow's reference was really a general one. Williams notes a marked increase in technical work during the 1850s which, he says, resembled the 1820s in its intensity.[41] In 1865, we find the old and infirm doyen of English physics still checking optical arrangements in lighthouses for Trinity House.

Faraday saw his work divided into two categories, viz. technical assignments done for the Institution, and personal research; but the former can be further subdivided in that much of it served Utilitarian purposes. As we have seen, Faraday did not have Brande's political awareness or professional goals, and made no such distinctions. Historically speaking, however, there is a world of difference between researches on glass or steel and the disinfection of prisons or the analysing of food for convict ships. By the mere fact of being in the position of Brande's assistant, or RI chemist, Faraday made a significant contribution to the Utilitarian case for the scientific 'solution' of social problems. We shall deal with both simply Institutional work, and the more Utilitarian researches, in turn.

As might be imagined, most of the strictly Institutional work done by Faraday was not very interesting. Such work was undertaken to demonstrate the industrial value of science, although many of the routine analyses were (during 1815–30) often medical in nature as well. Since fees went to Brande and Faraday rather than the Institution, the goal was presumably to build the RI's reputation by demonstrating the general economic utility of science. Most

---

[38] *LPW*, I, 181.
[39] *BJ*, II, 112.
[40] RI Archives, Box File IX, Folder 121C. The printed memo is entitled 'Report on the proposed School of Practical Chemistry'; the MS is entitled 'For a Practical Chemical School'.
[41] *LPW*, pp. 479ff.

of the assignments were immediate, day-to-day projects, but a few lasted several years. The manufacture of improved optical lenses between 1824 and 1830 was the most famous of these, undertaken at the instigation of the Admiralty. Ties between the Admiralty and the Royal Society went back many years, with the RS occasionally being funded for things like magnetic and geodetic surveys. But the RS was in no way capable of undertaking laboratory or industrial research, and thus when the Board of Longitude raised the issue of improving telescopes, the Society naturally turned to the RI. The project was officially in the hands of Faraday, John Herschel, and the optician George Dollond, but as might be expected, it was Faraday who bore the brunt of the work.[42]

In the early stages of the work the laboratory employed was the workshop of the glassmakers Pellatt and Green, but by September of 1827 this proved unfeasible and a furnace was constructed at the RI itself.[43] The technical difficulties were as follows. For optical purposes, glass must be homogeneous in composition and structure, otherwise different portions of the lens will possess different indices of refraction. Flint glass is optically the most desirable, but it presents real problems in terms of homogeneity. It contains lead oxide, so heavy that it sinks to the bottom of the mixture. Thus when Herschel received an early batch of specimens from Faraday in 1825, he wrote in response:

> The difference of S[pecific] G[ravity] between the top & bottom in one case exceeds anything I could have supposed possible. It is evident that no accidental defect in mixing could have produced it.[44]

There were other problems as well. The pots tended to crack and fuse under the heat, and ultimately the furnace itself began to go to pieces.[45] The glass contained bubbles and striae, which could not be removed by stirring the molten glass because this shattered the pots. Faraday wrote to Herschel that 'we shall have to feel our way a good deal',[46] and in fact the entire procedure was an excruciatingly tedious process of moving from problem to problem in an area that was without precedent. It was, as Williams notes, 'an exercise in precision cooking'.[47]

The notebook of this work is preserved at the RS, and is a rare example of an early industrial journal. The amount of scientific theory itself is almost non-existent. Rather, the notebook reveals a slow solving of problems by rational technique, i.e., painstakingly isolating the variable factor responsible for cracking, bubbles, striae, or whatever, and then altering the procedure. There is probably no better example of a merger of 'scholar and craftsman', though it

---

[42] The following discussion follows that given in *LPW*, pp. 116–20.

[43] *See* correspondence between Faraday and Herschel in the RI Archives, Box File XIVB, Folder 152.

[44] *LPW*, I, 149; dated 13 September 1825.

[45] ibid., p. 171; Faraday to Herschel, 24 July 1828.

[46] ibid., p. 152; 14 November 1825.

[47] *LPW*, p. 118.

took place within the same person.

The notion of an industrial laboratory in the heart of Mayfair is a bit amusing. Faraday himself was worried about the attention it might attract. Thus he recorded in the notebook late in 1827:

> The chimney was very hot even to the top too hot to be pleasant to the hand. On going upon the roof I observed sparks issue from it & on holding a piece of paper in the current of an issuing from it it was instantly charred sparks burn from it & it would have broke[n] into flame but I did not wish to attract the attention of our neighbours . . .. When the day darkened the sparkes [sic] from the flue became more evident[48].

In the end, Faraday obtained his lens, and results were read in the Bakerian Lecture to the RS on 19 November, 3 December, and 10 December of 1829.[49] The Royal Society's own report on the work, in 1831, found the lens powerful and very achromatic, 'indicating that all has been done that could be done by the artist who made it'.[50] They resolved

> That Mr Faraday be requested to make a perfect piece of Glass of the largest size that his present apparatus will admit. [A]lso that he be requested to teach some Person to manufacture the Glass for general sale.[51]

The latter request was never fufilled. Faraday may have demonstrated the advantages of scientific expertise and of a superior laboratory, and this undoubtedly impressed the Institution's governors; but as years of research and hundreds of pounds had gone to produce a single lens, the material effect on the business community was negligible. Added to that was Faraday's complete lack of interest in the work. As he told Dionysius Lardner in 1827, his commitment to the glass project was 'to a great extent for the sake of the Institution'.[52] As for continuing the work, he had this to say to Davies Gilbert, the new (in 1830) RS President:

> I further wish you most distinctly to understand that I regret I ever allowed myself to be named as one of the Committee. I have had in consequence several years of hard work; all the time I could spare from necessary duties (and which I wished to devote to original research) been [sic] consumed in the experiments and consequently given gratuitously to the public I should be very glad now to follow Mr Herschells [sic] example & return to the prosecution of my own views . . ..[53]

Faraday, in fact, regarded the project as a failure,[54] but it was something of a success in the tradition of Davy's work on agricultural chemistry. It received

[48] RS Archives, Faraday's glass notebook, p. 31, entry for 19 December 1827.
[49] Michael Faraday, 'On the manufacture of Glass for optical purposes', *Phil. Trans.*, CXX (1830), 1–57.
[50] RS Archives, Domestic MSS, Vol. III, Items 160 and 161.
[51] ibid.
[52] *LPW*, I, 169; 8 October 1827.
[53] ibid., p. 180; 13 May 1830.
[54] *BJ*, I, 402, and de la Rive, 'Michael Faraday', p. 416.

much publicity, in addition to the Bakerian Lecture and a discussion in Brande's journal. In 1831, for example, one scientific lecturer popularized the glass experiments as an example of the dependence of industry on pure science. 'The mere manufacturer,' he told his readers,

> . . . unenlightened by chemical knowledge, could never have arrived at a competent insight into the causes which produce striae, and which apparently must continue to produce them, in flint-glass. The chemist alone could have devised and executed the production of a new kind of glass . . ..[55]

He went on to predict a profitable market for the new lenses.

Although the project was economically unfeasible, the glass work helped to reinforce the common view of science as ancillary to industrial development. Glass technology is of immense complexity even today, and the 'science' Faraday brought to bear on the problem was largely a matter of trial-and-error and superb technique. Yet surely the RI was breaking new ground. The 'scientific' analysis of manufacturing processes may have been popular in the 1820s, but it was rarely carried out. The RI provided a controlled environment for a series of long, expensive, and laborious investigations. The economic failure became a social success as an advertisement of the advantages of a well-equipped laboratory devoted to technical needs.[56]

There are several sources of information on Faraday's technical researches, in addition to published items, such as *LPW*, I, II, and *Faraday's Diary*.[57] Extensive as this material is, it probably records less than half of Faraday's actual work.[58] Yet even the extant material reveals so dense a schedule of

---

[55] E. W. Brayley, Jr., *The Utility of the Knowledge of Nature Considered* (London: Baldwin and Cradock, 1831), p. 33.

[56] The work on steel with James Stodart, 1818–23, falls into the same category. *See* Robert A. Hadfield, *Faraday and his Metallurgical Researches* (London: Chapman and Hall, 1931), and his article, 'A Research on Faraday's "Steel and Alloys"', *Phil. Trans.*, Series A, CCXXX, 221–92; *QJS*, VII (1819), 288–90, and VIII (1820), 319–30; *Phil. Mag.*, LVI (1820), 26–35, and LX (1822), 335–63 and 363–74.

[57] Thomas Martin (ed.), *Faraday's Diary* (8 vols.; London: G. Bell and Sons, 1932–36). Unpublished items include a series of letters from Faraday to J. W. Lubbock (RS Archives, MSS. LUB. F.3–F.6), discussing Excise work during 1835; letters from I. K. and M. I. Brunel, on hospital ventilation and the Thames Tunnel construction, at the Institution of Electrical Engineers; the RI laboratory notebooks (four volumes, extending over 1809–59; a manuscript volume entitled 'Faraday MS 11', also at the RI, containing notes made during 1827–50; nine Letter Boxes of Faraday correspondence (RI Archives, FL-1 to FL-9); the Magrath letters in the Dawson Turner Collection at the Library of Trinity College, Cambridge; and a collection of correspondence (a few items of which have been published), at the Library of the Wellcome Institute of the History of Medicine (File 69293).

[58] A random examination made by Alan Jeffreys of Admiralty Papers at the Public Record Office for 1830–5 turned up three letters by Faraday, and of course Faraday did extensive work for a number of government departments (*see* Alan E. Jeffreys, *Michael Faraday: A List of His Lectures and Published Writings* [London: Chapman and Hall, 1960], p. xxii. Cf. also *LPW*, I, 508: in a letter to Lord Auckland, Faraday cites requests from the Home Office, Admiralty, Ordnance, Department of Woods and Forests, and 'others'). Secondly, Faraday published a number of unsigned articles in Brande's journal (most, however, identified by Jeffreys), and we can imagine that there was much more work which he did not bother to report (after 1831, of course, there was no journal). In a number of cases his correspondence makes oblique references to chores that were never recorded. Finally, the records of his work for Trinity House—nineteen folio volumes—were destroyed by bombing during the Second World War.

technical researches during Faraday's lifetime that one can only wonder how he managed to postpone a nervous breakdown to 1839. The sources depict a man constantly on the move all over England, testing lighthouse arrangements, checking gas works, or visiting manufactories; a man whose word of approval had become, by the 1830s, the *sine qua non* in technical matters. Marc Brunel wrote to Faraday regarding his recent reply to a query on the Thames Tunnel construction:

> I might take upon myself to report what you have said to me upon this question; but a word or a line from your Pen will carry much greater weight among all our interested friends—and most particularly with those who lend us the money to go on with the work.[59]

The material creates, finally, the impression of the RI Laboratory as a scientific powerhouse, bearing little relation to the genteel outward appearance of the place. Virtually every imaginable type of analysis is recorded in Faraday's notebook: heating fungus to obtain ammonia, analysing catechu fluid for a solid week, examining tar, tea, linen, wines, paper, cow medicine, bark, blood, apple juice, coal, cheese, nitre, malt, dye, lacquer, asparagus, flour—there is no end to the list. In addition to these random analyses, a large part of the work came from long-term appointments, even beyond the work for Brande done during the 1820s. Some of this had a distinct military flavour. Faraday taught at the Woolwich Military Academy, 1829–49; was appointed Scientific Adviser to the Admiralty in 1829 (although this began unofficially in 1825); and was similarly appointed to Trinity House in 1836, a job which proved to be enormously time-consuming. It would be both unnecessary and tedious to detail all of this, and I cite it to demonstrate how the understandable interest of intellectual historians in the electromagnetic researches has all but obscured Faraday's powerful 'Mr Fixit' image that prevailed from 1830 to 1860.

It is difficult to say to what extent Faraday's Institutional work was Utilitarian in nature or inspiration. Certainly the major efforts of the 1820s were (excepting the steel and glass research), but these arose out of directives from Brande. In terms of actual work time, much of the Institutional work consisted of routine commercial analyses. Yet some of the projects were clearly congruent with Utilitarian ideals, and the cost analysis technique which Faraday had learned at Brande's side was the methodology of efficiency that was so much a part of Bentham's vision. The use of Faraday by the government need not, of course, have been Utilitarian in nature; and many Utilitarian plans—Chadwick's hopes for sanitary reform, for example—suffered serious opposition and/or defeat. Yet the scale on which Faraday's services were employed, the types of problems that had come to be classified as scientific, and the nature of his approach to them, seem to be something of a break with the past. I am not, in short, trying to demonstrate that science was fully Utilitarian by the 1840s, which was obviously not the case. But it does seem clear that we

---

[59] IEE, Faraday Papers, dated 11 February 1836. *See* also *LPW*, I, 342–3.

are, during the 1830s and 1840s, witnessing the beginnings of a trend. By that time, the Royal Institution hardly had to make appeals to the government, as it did in 1809, to be of service. Instead, the government had a scientist (and not just Faraday) 'on tap', and Faraday was deluged with requests for jobs that were formerly not associated with scientific research. By 1844, it was apparently regarded as perfectly logical to appoint a scientist to investigate a mine disaster.

To take some specific examples, cleanliness and disinfection were the type of projects that interested Utilitarians, for reasons that went far beyond public health. The association of dirt with chaos is well known to psychologists and anthropologists,[60] and a personality such as Chadwick's typified this fear. A work such as his *Report on the sanitary condition of the labouring population of Great Britain* (1842) exemplified the attitude that if only the country were neat and tidy, there would be little to worry about. We have already made mention of Faraday's career as a water analyst, and his disinfectant work was also a part of this world view. When a fever epidemic broke out at the Milbank Penitentiary in February 1823, a Select Committee was appointed to investigate. Commissioners included four governors of the RI; and the four investigating physicians were Roget, Granville, Gilbert Blane (all RI governors) plus P. M. Latham, Faraday's personal physician.[61] Faraday did some analyses for Milbank in June, and the following year was appointed to fumigate the prison, on the suggestion of Davy, then President of the Royal Society.[62] Using liquid chlorine, Faraday was apparently successful, and wrote a description of the assignment for Brande's journal, giving his ideas on the general subject of the purification of buildings.[63] From that point on, disinfectant work became a part of the Royal Institution's repertoire, and Faraday the recognized expert. The RI arranged a Friday Evening Discourse for 2 February 1827, at which a Mr Alcock discussed the use of disinfectants in police work and in work done on sewers.[64] Faraday, who had done a study of Labarraque's disinfectant, followed up Alcock's Discourse with one of his own, on 4 May 1827, discussing the chemical basis of disinfection.[65] Faraday may have even made his own mixture for prison and police use. Edward Magrath wrote to his friend Dawson Turner in 1831 of

---

[60] For example, *see* Mary Douglas, *Purity and Danger* (New York: Praeger, 1966).

[61] Report from the Select Committee on the State of the Penitentiary at Milbank. Ordered, by the House of Commons, to be Printed, 8 July 1823. The RI governors who were commissioners were W. M. Pitt, Giles Templeman, Davies Gilbert, and Benjamin Hobhouse. Milbank was designed on the basis of Bentham's model prison, the 'panopticon'.

[62] Report from the Select Committee on the Penitentiary at Milbank. Ordered, by the House of Commons, to be Printed, 11 June 1824. On Faraday's 1823 analysis *see* RI Archives, Laboratory Notebook, Vol. 1 (1821–29), entry for 3 June 1823.

[63] 'On Fumigation', *QJS*, XVIII (1825), 92–5. Faraday records in his Laboratory Notebook, Vol. 2 (1830–59), sending alkali to a penitentiary, possibly Milbank, in January 1830.

[64] 'An Account of the Extensive Applications lately made in Paris of the Chlorides or Chlorurets of Lime and Soda, as disinfecting agents', *QJS*, n.s., I (1827), 211.

[65] 'Experiments on the Nature of Labarraque's disinfecting Soda Liquid', *QJS*, n.s., II (1827), 84–92; and *see* I (1827), 460–2.

USEFUL
To Prevent Infection.

Chloride of Lime,---Bleaching Powder, 6d. per lb.

TO BE HAD OF

Mr. Henry BARNES, Dry Salter, 110, Long Acre,

OR,

99, Upper Thames Street.

To Make Liquid.

Suitable for Use in Schools, Workhouses, Workshops, Police Offices, Theatres, &c. &c.

Put 1lb. of Good Powder into a Glazed Earthen Vessel, add Five Quarts of Water; Stir it well.---let it remain an Hour,---Pour it off into Bottles for Use.---It need not be Strained. By adding a Quart of Water, a Sixth Quart of Liquor of sufficient strength will be obtained.

MANNER of USING the LIQUID.

1. Soak a coarse Cloth in it and suspend the Cloth in a room after persons have left it. Let it hang, and renew it every 24 Hours.---Sprinkling the Floors is sometimes adopted, but it is injurious to them.

For Use in Bed Rooms.

2. Put Two ounces of the Dry Powder in a Plate or Saucer in any elevated safe place,---Stir it occasionally, and it will last a Fortnight or Three Weeks.

For Use in Stronger Cases of Foul Air.

Put a Large Table Spoonful of the Liquor into a Pint and a Half of Water; also, a similar Spoonful of Good Vinegar,---mix, and place, the Mixture in shallow dishes in the Room.

For Use where the Air is in a VERY BAD State.

One Ounce of Powder, Half a Pint of Water, and a Tea Spoonful of Oil of Vitriol. Avoid the Fumes.

N.B. Chlorine Gas being very Destructive to Metal, bright Fire Irons, and other Household utensils should be kept from its action.

The quantities mentioned can only be general; experience will point out what ought to be used in proportion to the size of the Room, &c.

1. This is a Useful Process where many persons assemble in the DAY.
2. This Process has been used by a well-known Chemist, in his own Bed Room, for the last Five years.

SAYERS & SON, Printers, 67, East Street, Marylebone.

23. 'Useful to Prevent Infection', advertisement for bleaching powder, possibly prepared by Faraday

24. Faraday as a young man

25. Faraday in later years

26. A Christmas lecture at the RI, 1849, before the Prince Consort and the Prince of Wales

Faraday giving his Card to Father Thames

(Courtesy of the Managers of *Punch*)

27. *Punch* cartoon of Faraday, 1885

28. Davy's safety lamps

a new process which Mr Faraday [regards] as particularly applicable to Police Offices and other places when filled with crowds of People. He recommended it strongly to his Pupils at his Lecture yesterday.[66]

The next day, Magrath sent Turner a printed notice (*see* Plate 23) advertising the bleaching powder mix, noting in the margin its usefulness for Police Offices (at this point Peel's Metropolitan Police was only two years old).[67] One can only wonder whether the anonymous 'well-known Chemist' who used it 'in his own Bed Room, for the last Five years', was not Faraday himself.[68]

Faraday apparently continued in this sort of consultant capacity for many years. When the First Lord of the Admiralty wrote to him on the subject in 1847, Faraday replied that his mental condition had forced him to give up all but theoretical research (which was certainly untrue). 'If I were in London,' he told Lord Auckland, 'I would wait upon your Lordship, and say all I could upon the subject of the disinfecting fluids, but I would not undertake the experimental investigation ...'.[69] During the 1820s, however, such investigations were commonplace.

Faraday's government work, such as lighthouse maintenance, was not necessarily specifically Utilitarian, but it usually involved cost-benefit analysis. Directly after his appointment to Trinity House, Faraday began drawing up tables of oxygen manufacture for the light at Bude. How much did it cost to produce a given illumination intensity? Faraday wrote in his notebook: 'exactly 1.909 pence'.[70] Yet equally important as this Benthamite calculus was the *fact* of scientific management of such problems by the government, which Faraday's own life represented. Thus R. K. Webb writes of sanitary engineering (for example) that 'the experts disagreed violently about the best technical solutions to a given problem',[71] but omits the more significant point that there was no disagreement that the solution (whatever it was) should be technical. This attitude was beginning to spread to every phase of social life.

Faraday viewed all this as a distraction from his 'real' work, yet regarded it necessary for a loyal subject to provide government service. In this sense, he was not apolitical, but very political—a Tory, as Williams notes.[72] The anagrams Faraday played with in his *Commonplace Book* when in his twenties strongly bear this out:[73]

[66] Magrath to Turner, 14 December 1831, Trinity College Library, Dawson Turner Collection, Class Mark 0/14/7, p. 206. Citation by permission of the Master and Fellows of Trinity College.

[67] ibid., letter of 15 December 1831, p. 207.

[68] Magrath's letter of the 14th December seems to imply this, as it begins: 'I have not yet received the printed copies of the enclosed which Mr Far[MS torn] allowed Mr Haskins to have executed—'.

[69] *LPW*, I, 508.

[70] *BJ*, II, 87.

[71] R. K. Webb, *Modern England* (New York: Dodd, Mead, 1968), p. 259.

[72] *LPW*, pp. 357–8. *See also* pp. 41–2, and *LPW*, II, 558.

[73] Kept in the archives of the IEE. Examples are taken from pp. 391, 404, and 434. The idea, of course, is to rearrange the letters to form another word or phrase, facetiously taken to be synonymous.

Democratical—comical trade
Revolution—to love ruin
Old England—golden land
Sovereignty —'Tis ye govern
Radical Reform—Rare Mad Frolic
Universal Suffrage—guess a fearful ruin
Monarch—march on
Sir Francis Burdett—Frantic Disturber
William Cobbett—I'll be at the mob W.C.

Faraday had absorbed all of the prejudices of the upper classes, for whom, no doubt, he was living proof that a hard-working young man could, à la Samuel Smiles, make it in the system. His scientific interest, after all, had been kindled by reading a diffusion-of-knowledge book on chemistry, written by the wife of an RI governor, Jane Marcet, who also wrote a popular treatise on political economy. Starting from this early self-education he had elevated himself by sheer hard work (and a lucky break from Davy) into a position of great eminence. 'The story of Faraday,' exclaimed the RI President in a letter to Charles Babbage,

> is just one that is sure to make a great noise. There is something romantic and quite affecting in such a conjunction of Poverty and Passion for Science, and with this and his brilliant success he comes out as the Hero of Chemistry.[74]

Faraday was a stern taskmaster with his subordinates (one of whom threatened to kill him after being fired[75]), and found habits such as idleness, begging, or drinking repulsive.[76] 'Your words repel me,' he wrote in reply to a beggar's request.

> [Y]our flattery . . . is ever hateful—I cannot give the half crown you ask for—It will leave my hands because you ask for it but it will go to the poor, who are not destitute by their own acts of intemperance.[77]

Just as Faraday regarded scientific controversy as destructive, so did his class background, and his religious temperament, lead him to see any sort of social disturbance as evil. He did not view his public service as social engineering, of course, but the 'law and order' mentality was hardly foreign to his own political outlook. The use of science for such purposes received its finest expression in Faraday's investigation of the explosion at Haswell Colliery (recounted in some detail below); and as such scientific commissions on social problems have become the order of the day, an examination of the event has much to tell us about the role of science in industrial society.

To anticipate for a moment, the result of Faraday's investigation[78] was that

---

[74] BM Add. MS. 37189, f. 218, the Duke of Somerset to Charles Babbage, 16 December 1835.
[75] *MM*, VIII, 329–30, 379–80, 387, and 389–90.
[76] *See* letter to Dr John Forbes, 29 December 1841, at the Linnean Society.
[77] RI Archives, Letter Box FL-9.
[78] The investigation was theoretically the joint work of Faraday and Sir Charles Lyell, it being

no one was to blame, and that the disaster was an accident. In other words, there was no evidence of negligence, of by-passing the customary safety precautions that were normally in effect. What these precautions were, however, was never even mentioned, nor was it ever made clear what would *not* constitute an accident. Within such a narrow framework, the inquest would necessarily have to ignore the larger issue of the nature of mining itself, or to what extent safety regulations were drawn up so as to allow increasingly deeper (and thus more hazardous) penetration of the earth. 'No matter whose system [of mine ventilation],' James Ryan, a Director of Mines, testified in 1835, 'if man and gas travel together, there must be calamites.'[79] Although this was not deliberate, the legal ideology of science had precisely the function of focusing on the narrow technical issue, so that the nature of a society that produced mine disasters was never called into question. In this regard, the Select Committee to which Ryan testified may have been somewhat unusual, for the report noted that the Committee

> would gladly put it to the owners of these Mines how far any object of pecuniary interest or personal gain, or even the assumed advantages of public competition, can justify the continued exposure of men and boys in situations where science and mechanical skill failed in providing anything like adequate protection.[80]

Few, however, recognized that the cause of mine disasters was not simply a dysfunction in the Davy lamp, but an industrial economy based on profit and production. In Faraday's hands, the function of science was not to explain the disaster, but to explain it away.

The major breakthrough in mining safety was the Davy lamp of 1816, in the research on which Faraday had, in fact, assisted.[81] Yet the discovery of wire gauze as a means of cooling the flame resulted in more, rather than fewer, accidents. According to the 1835 report, there were 447 deaths due to explosions in Durham and Northumberland during 1798–1816, and 538 deaths in the equivalent period after the introduction of the lamp, i.e. 1816–34. The reason for this was the use of the lamp to work deeper and more dangerous seams, in conditions of too high temperatures and improper ventilation.[82] Although it had been tested at Wallsend and other collieries, most of the experiments were done in the Laboratory, and as Davy himself pointed out, the lamp had crucial limitations. A rapid air current, or very hot surroundings, could render the

presumed that geology was relevant to mining. In actual fact, the work was carried out and written up by Faraday, as Lyell himself admitted.

[79] Report from the Select Committee on Accidents in Mines . . .. Ordered, by the House of Commons, to be Printed, 4 September 1835. James Ryan was a director in Montgomeryshire, and his testimony was heard on 6 July 1835. Ryan's plan for draining gas from mines had been adopted to some extent in 1808, and one RI governor, Lord Dudley and Ward, was among the earliest to employ the system in his collieries.

[80] ibid.

[81] On the lamp *see* Harold Hartley, *Humphry Davy* (London: Thomas Nelson and Sons, 1966), pp. 109–21.

[82] J. L. and Barbara Hammond, *The Town Labourer* (Garden City: Doubleday, 1968 reprint), pp. 22–3.

flame dangerous.[83] Faraday stated at the Haswell inquest that Davy's instructions had limited value:

> Sir Humphry was not infallible. He was not a practical miner, and therefore such instructions could only be a matter of opinion. [However] A large portion of the coal now worked could not have been worked had safety lamps not been invented.[84]

The last sentence reveals the reason for the increased number of disasters, for Faraday was inadvertently stating that the invention of the lamp had essentially made possible greater profits, not greater safety.

A large portion of the 1835 report was devoted to testimony on the Davy lamp. The Committee, it should be added, contained two RI governors (Lord John Russell and David Barclay), as well as two mine owners who had close ties with the RI and had used its services (J. H. Vivian and J. J. Guest). What emerged in the course of the extensive testimony was not merely the revelation of an increase in accidents after the introduction of the lamp, but that the lamp (as Davy had pointed out) was not failsafe. Thus Goldsworth Gurney, who described himself as a man devoted to mechanical and chemical pursuits, opened the inquiry by testifying that the Davy lamp 'is not pratically safe; experimentally it is so ... but in practice I do not think it is'.[85] James Ryan testified that a Mr Allen (probably William Allen) had asked him to test the lamp in 1817, and he discovered that it was easily exploded by creating a draught. Ryan's testimony is so striking, and so revealing of the problem of shadow vs. substance in the role of science in the Industrial Revolution, that it is worth quoting at length:

> I went to the gas works in Peter-street; I exploded it there; and I was then challenged in the matter, and in the Society of Arts' rooms, before a large committee in that house. I exploded it with throwing coal dust at it; with changing the current of air, and taking and putting it there; or taking it to the door, and shutting it. I will explode it by that; I will explode it by my coat, or I will explode it by my hat. Anything that would produce a current. It is a very extraordinary fact that this never appeared in the Transactions of that Society.
>
> 2940. But are you not aware that Sir Humphrey [*sic*] Davy gave positive directions to all those who were likely to avail themselves of this invention, never to allow the lamp, on any consideration whatever, to be exposed to a sudden current?—I never heard that. I heard Sir Humphrey Davy on the thing, and I never heard him say that. I wished he had published that, it would have saved many.[86]

[83] Hartley, *Humphry Davy*, p. 116.

[84] *The Times*, 14 October 1844, p. 2.

[85] At one time Gurney thought he had found a way around this by using a system of reflecting mirrors to illuminate mines, but the plan was considered impractical (*The Times*, 11 October 1844).

[86] Report from the Select Committee on Accidents in Mines ..., 4 September 1835, testimony of the 6th July. Four days later, John Roberts, a lamp manufacturer, also said that the problem of currents was never mentioned to the miners. It must be reiterated that Davy did publish on the subject. He read a paper before the Royal Society in January of 1817 (there were no miners present), and published *The Safety Lamp for Coal Mines with some Researches on Flame* in 1818.

The Davy lamp, then, was unreliable in certain practical circumstances. Since mines had to be ventilated to keep the gases from causing suffocation, they were subject to sudden air currents, and this intake of air also formed explosive mixtures. Even as minor an accident as denting by a fall, or damage by a rock, rendered the lamp a major hazard. The Davy lamp was reliable when the air in a mine was perfectly still, but when in motion, said George Birbeck on the 20th July to the Committee, 'it becomes no longer a safety-lamp; the flame within then passes through the meshes and ignites the explosive compound that is without'.[87] The situation was simply too variable. Mines vary radically in air flow and shape, and conditions are never identical from day to day even in the same mine. Yet although the 1835 report came very near to discrediting the practical deployment of the Davy lamp, it never suggested that there was a limit to the scientific approach to such problems. Science, in fact, emerged a hero. Throughout the report is the repeated suggestion of scientific investigations and scientific education as the solution to mine accidents. Mechanics' institutes, polytechnic schools, commissions of scientific men, etc., were advocated again and again.[88] And in 1844, when the Home Secretary's assistant wrote to the coroner at Haswell to inform him of Faraday's and Lyell's impending arrival at the inquest, he said that the Secretary decided to use a scientific team because of the recommendations of the 1835 report.

To return to the events at Haswell, besides the official report and subsequent publicity, there are very few sources for the story of Faraday's investigation. *The Times* gave the coroner's inquest major coverage—a large number of columns spread out over October 2, 3, 4, 11, and 14 of 1844—but only the last two days of this discussed (slightly) Faraday's role. Faraday also took some notes at the inquest (preserved at the IEE), and the RI has in its possession a long letter from Lyell to Bence Jones, written in 1868, describing the affair from his memory as best he could.[89] From these sources the following story can be reconstructed.

On 28 September 1844 an explosion at the Haswell Colliery, in the centre of the Durham coal fields, caused the deaths of ninety-five men and boys. At the Home Secretary's request, Faraday and Lyell travelled to Durham on October 8th, and proceeded to Haswell the next day. They attended each session of the inquest and, on October 10th, went into the pits for eight hours, to examine the mine and talk to the men. The following day the jury ended its examination of witnesses and brought in a verdict of accidental death with which, Faraday wrote in his notes, he completely concurred.[90] The following day, he and Lyell returned to London, and on 17 January 1845 Faraday delivered a Friday Evening Discourse on the affair which attracted wide attention.[91] Colliery explosions

[87] ibid., testimony of the 20 July.
[88] ibid., summary, as well as Birkbeck's testimony (20 July), Jonathan Pereira's (21 July), *et al.*
[89] RI Archives, Letter Box FL-8, dated 22 May 1868.
[90] IEE, Faraday Papers, MS notes of the investigation; *BJ*, II, 185.
[91] Besides newspaper reportage of the lecture, it was reprinted verbatim in *The Repertory of Patent Inventions*, enl. ser., V (1845), 191–211, and *The Civil Engineer and Architect's Journal*, VIII (1845), 115–18.

were also discussed at the 1845 meeting of the British Association, and Faraday contributed to the debate at the Section C meeting, Geology and Physical Geography.[92] Their final report for the government was printed by the House of Commons on 18 April 1845, and reprinted in *The Philosophical Magazine*.[93]

The Haswell explosion was an especially touchy subject for two reasons. First, there had been a strike at Haswell recently, and class feeling was still running high. If the investigation could have revealed negligence on the part of the owners, workers' opposition, already in evidence, could have gotten out of hand. The issue was, Lyell wrote Bence Jones, 'whether any undue considerations of economy had induced the owners of the mine to neglect such precautions as are customary & indispensable to the safety of the men employed in the work'. [94] Secondly, Haswell was, from the point of view of safety, a model mine. Again and again at the coroner's inquest workmen and viewers alike testified that they had never been in a mine so excellently ventilated. Thus Faraday wrote in his 1845 report: 'The mine has the character of being one of the best ventilated in the whole trade, a circumstance which ... makes us more anxious, if possible, to discover [the disaster's] cause ...'. Hence if Haswell could have been the scene of such a disaster, other mines in the north of England were even more liable to such a calamity.

To return to the actual investigation, it can be divided into two parts, namely the inquest itself, and Faraday's day-long visit to the mine on October 10th. If the reporting in *The Times* is to be believed, the inquest was something of a whitewash. The miners' representative, a Mr Roberts, was labelled 'the pitmen's attorney-general', and was repeatedly denied the right to have the mine examined by viewers chosen by the workers themselves. These men were James Mather and Matthias Dunn, who the owners' solicitor, Mr Marshall, claimed were biased towards the workmen. Roberts argued that he 'knew of no other viewer [besides Dunn] in whom the pitmen had confidence', but Marhsall said his employers had a personal objection to him.[95] The argument on this point became quite heated, and on October 4th Roberts was nearly ejected from the court by the coroner over the issue. At the end of that session, he again made his request, and again was turned down.

As the testimony proceeded, it became evident that there was very little possibility of determining the cause of the disaster. Three witnesses who were present just before the explosion occurred testified that ventilation was excellent and that a candle test gave no indication of gas present. There was also agreement that the 'goaf', or worked-out area, was sufficiently ventilated

[92] *The Athenaeum*, 26 July 1845, p. 746.

[93] Copy of the Report of Messrs Lyell and Faraday to the Secretary of State for the Home Department, on the subject of the Explosion at the Haswell collieries in September last:—Also, Copy of the Report Addressed to the United Committee of the Coal Trade by the Special Committee appointed to take into consideration the Report of Messrs Lyell and Faraday; and Copy of the Reply of Messrs Lyell and Faraday thereto. Ordered, by the House of Commons, to be Printed, 18 April 1845; and *Phil. Mag.*, XXVI (1845), 16–35.

[94] RI Archives, Letter Box FL-8, Lyell to Jones.

[95] *The Times*, 2 October 1844.

so as to prevent an accumulation of gas there—an important point, because this was one area in which gas was usually trapped, which could then be fired by a damaged Davy lamp. Thomas Forster, a Haswell viewer, was particularly emphatic on this point. 'I know of nothing,' he added,

> that could have been done to prevent this explosion . . . . It is my opinion that by an investigation of the mine by scientific and practical men the cause of the explosion could not be obtained with certainty.[96]

In the last two days, however, the nature of the evidence changed.[97] Two outside viewers, not connected with Haswell but hired by the owners, had made an examination and reported their findings. Nicholas Wood, a colliery viewer at Killingworth, asserted that the upper regions of the goaf may have been poorly ventilated, and that the explosion was caused by an accumulation of gas in that region. George Hunter, colliery viewer for the Marquis of Londonderry, said that the cause was due to a damaged lamp plus gas in the goaf.[98] The most curious testimony was given by one of Lyell's friends, Samuel Scutchbury, a surveying agent to the Somerset district of mines held by the Prince of Wales, who, it should be added, went down with Faraday and Lyell into the mine on October 10th. Scutchbury stated, remarkably enough, that he had 'formed my opinion before I had obtained any knowledge of the circumstances of the explosion', and that the testimony given up to now served to confirm it, viz. that the cause was gas in the goaf.[99]

At this point, the coroner moved to clear the court, but Roberts objected that he had not yet produced witnesses on behalf of the workmen, and wished to do so. The coroner, however, 'did not think that necessary now, and again ordered the court to be cleared . . . .'[100] After a few minutes' deliberation, the jury brought in a no-fault verdict, and the trial, if it can be so called, was over. According to Lyell (in the letter to Bence Jones) there was also conflicting testimony from the proprietors and the pitmen over the nature of the safety precautions that were in effect, but this second sensitive issue was also successfully buried.

Faraday's role in this affair amounted to giving the decision the status of scientific repectability. He not only agreed with the verdict, but with the supposed cause of the accident, which had been stated as a matter of opinion by

[96] ibid., 4 October 1844.
[97] This is discussed in Lyell's letter to Jones.
[98] *The Times*, 11 and 14 October 1844.
[99] ibid., 14 October. It is interesting to note that Scutchbury had at one time been in the personal employ of RI governor I. L. Goldsmid. Correspondence between William Henry Fitton, from the Geological Society, to Goldsmid, which is in the second volume of a two-volume collection of Goldsmid letters kept at University College, London, reveals this association (*see* letters of 28 June and 10 July 1828). Scutchbury apparently stayed at the Goldsmid house to do his work; Fitton wrote Goldsmid to ask if he could 'borrow' Scutchbury to arrange the collections of the Geological Society. (It should be added that Fitton spelled the name 'Stuchbury' and 'Stutchbury', but it is undoubtedly the same person, as he is identified as a man possessing good knowledge of localities and field work.)
[100] ibid.

Wood, Hunter, and Scutchbury, and which had no scientific basis whatsoever. Nor was the Home Secretary unaware of what role Faraday's imprimatur would play. In the midst of the trial, his assistant wrote to the coroner at Haswell that as a result of that imprimatur, 'to the Public (particularly to the Relations of the Deceased) the verdict would be delivered under the best possible recommendation and with the highest sanction'.[101] It is a rather grim comment. Even if science could not produce any answers, it would silence embarrassing questions.

If there was nothing scientific in Faraday's analysis of the accident, there was little that was original in his solution, a plan for ventilation of mine goaves which was a modification of a design given in the 1835 report by James Ryan. Yet the government was delighted with the results. The Home Secretary contacted Faraday after the inquest, asking him if he and Lyell could tour all the principal collieries in Durham and Northumberland.[102] Faraday replied that he thought it unnecessary: 'We think that we see the cause of nearly all the explosions in coal mines [!] & we think that the means which we propose for their avoidance are practically and immediately available'.[103] Mining inspector H. S. Tremenheere—'another of the new breed of civil servants'[104]—wrote to Faraday that there 'can be no doubt that you have hit upon the cause of the evil, and its remedy', and he told Faraday he would help him in the distribution of his report to mine owners.[105]

According to Lyell, Faraday was in full command of the situation both at the inquest and in the pits. He 'began after a few minutes,' wrote Lyell, 'to cross-examine the witnesses with as much tact, skill & self possession as if he had been an old practitioner at the Bar'.[106] As for their day in the pits, the only information they managed to obtain, according to Lyell, was that the miners were illiterate and poorly educated, a point made in their final report. Few of them, said Lyell, could write, and he thought danger in mines could be averted if miners would study chemistry, pneumatics, and geology.[107] In the notes for his Discourse of 17 January 1845, Faraday wrote: 'doubtful character of evidence from coal miners &c ... again coal mines are out of society's thoroughfare & criticism so improve slowly only—'.[108] The report also pointed out that France and Germany had large schools of mines for the purpose of practical education. Faraday emphasized this again at the British Association

[101] IEE, Faraday Papers, S. M. Phillipps to Thomas Maynard, 7 October 1844.

[102] ibid., S. M. Phillipps to Lyell and Faraday, 19 October 1844.

[103] ibid., draft reply by Faraday, 21 October 1844. A letter from Phillipps to Faraday on October 23 acknowledges receipt and notes that his letter was placed on the desk of Sir James Graham, the Home Secretary.

[104] Webb, *Modern England*, p. 258.

[105] IEE, Faraday Papers, dated 14 November 1844.

[106] RI Archives, Letter Box FL-8, Lyell to Jones.

[107] Yet Lyell mentions that the men devised a method of calculating the rate of air flow in the mine by dropping a few grains of gunpowder into a candle and noting the time it took the resulting cloud of smoke to be carried a certain number of feet. This figure was then converted into miles per hour.

[108] RI Archives, 'Faraday Coll^n Rough Lecture Notes 1838–62' (Box no. 61).

meeting the following year. 'The great source of the danger,' he told his audience, 'was the mental condition of the miners. With regard to the present race this was so hopeless, that nothing could be done for them . . ..'[109]

Faraday's report to the Home Secretary, which was published in April of 1845, was delivered on 21 October 1844. As we shall see, the mine owners replied on 27 March 1845 (both Faraday's report and the owners' replies were printed with the main report). On the same day on which the mine owners replied, Faraday received a remarkable piece of evidence from Thomas Maynard, the coroner, which rendered the verdict of the inquest, and Faraday's report, virtually arbitrary. According to Maynard, sometime after the incident and inquest the pitmen approached the Haswell viewer, Thomas Forster, saying that they were afraid for their safety. Forster asked them to appoint a committee of four to examine the pit, and they methodically went through all the goaves, some as high as eighteen feet, '*but could not in any part find the least indications of Gas—*'.[110] Maynard went on to say that this corroborated the testimony given at the inquest that no gas had been observed in the goaf. Another man, a Mr Taylor (not the mine owner), was asked to check this, and could not find any gas there either, much to his incredulity. If this was so, Maynard continued, then gas was only given out occasionally, and in large amounts. The implication, of course, was clear: the supposed cause of the explosion, far from being certain, was completely a matter of speculation. It *could* have been the cause, but it was just as likely that it was not. Maynard undoubtedly realized the spurious nature of his own inquest, for it was he who refused to let Roberts bring in spokesmen for the workmen, and this refusal had allowed a dubious conclusion to be rammed through unchallenged. He concluded his letter by urging Faraday to return to Haswell for a further investigation, which Faraday did not, however, undertake.

Faraday was not aware of these developments in mid-January, when he delivered his Friday Evening Discourse on the subject. From the reports, the Discourse, given before a record crowd of 643 people (more than twice the average attendance) must have been a lovely bit of entertainment.[111] Indeed, Faraday used the word 'beautiful' several times and assisted his presentation with numerous demonstrations, maps, models, diagrams, and blackboards. To pave the way for the presentation, he began with the following words:

> I may state summarily, that the conclusion we came to was, that the inquest was conducted most fairly, openly, and in a very enlarged manner; that, under the circumstances, the catastrophe was purely accidental. Ninety-five men and boys, it is true, were unhappily killed, but no fault could be found with the proceedings of the persons concerned in the management and working of the mine . . ..[112]

[109] Reported in *The Athenaeum*, 26 July 1845, p. 746.
[110] IEE, Faraday Papers, Maynard to Faraday, 7 February 1845. Italics in original; the words 'in any part' are underlined twice.
[111] RI Archives, Box File VII, Item 108, *Index for F. E. Meetings 1842–65*, entry for 17 January 1845.
[112] *The Civil Engineer and Architect's Journal*, VIII (1845), 115.

By removing the issue from the arena of politics, Faraday was then free to interest his audience in the technical details of the problem. He explained how a coal mine works, and the meaning of terms such as 'seam' and 'pillar'. He took a long time describing a goaf and its formation, and demonstrating (with the aid of an inverted oblong vessel as a model) how fire-damp (methane) rises into it and is easily exploded when it mixes with air in a ratio ranging from 1 : 5 to 1 : 14.[113] He also performed numerous mini-explosions. 'There can be no doubt,' he told his audience,

> that the goaf is a very convenient place for the gas to accumulate in; that there is a tendency to that in the goaf from ventilation, the arrangement of which I must not go into, but it is very beautiful ....[114]

Finally, he gave his own recommendation, viz. laying a pipe between the goaf and the upcast shaft, and drawing off the accumulated gas with a draught. 'The principle,' he told the governors and Members, 'cannot fail to answer.'[115] Faraday concluded by admitting that he had been anticipated by James Ryan, and stated that the 'principle of withdrawing gas ... is not new among the coal-owners'.[116]

The major impact of the demonstration lay in its style. Faraday and Lyell had discovered nothing, and what 'science' had to do with their investigation—except in the Utilitarian sense of the term—remains unclear. As to the cause of the accident, they simply adopted the testimony of Wood, Hunter, and Scutchbury; and as to the remedy, Faraday had just suggested a modified version of James Ryan's method. Ryan himself got the idea after patenting an invention for land drainage, and in fact had lectured upon it at the RI. But the value of Faraday's Discourse and its attendant publicity was the dramatic confirmation of the legal ideology of science. Affairs of this sort were apparently amenable to direct scientific control. They were *manageable.* Faraday's plan was never put into practice, but the import of the entire inquiry was that a highly charged political issue could be neatly converted into a 'technical difficulty'. Here was a mine disaster in an area recently torn apart by a strike, a disaster that could have triggered off another strike or even a riot. Instead, it had been gently subdued by a technological fix that was actually inadequate. The legal ideology of science was ultimately the ideology of containment.

The work described above formed the essential part of the 1845 House of Commons report, which explained the problem of the goaf and the need to drain it, and emphasized the importance of educating miners in chemistry, pneumatics, hydrostatics, and geology. In this respect, it contained no surprises. Yet one item did emerge, almost inadvertently, that pointed to what was

---

[113] Davy's figure for the lower limit, according to the 1845 report, was 1 : 16. Faraday actually used coal gas in the Discourse.

[114] *The Civil Engineer*, p. 117.

[115] ibid.

[116] ibid., p. 118.

apparently the real cause of such accidents, though Faraday dropped the matter after referring to it in passing. He began the report by exonerating the colliery owners, and then on the following page described the prevalence of a practice that has certain affinities to today's practice of strip mining. The owners pursued the policy of leaving upper seams of good coal untouched 'in order that the thicker and more valuable seam . . . should first be worked out. In adopting this plan', continued Faraday,

> the proprietors have been guided by considerations of present profit, which the competition of other neighbouring coal works renders indispensable. Nevertheless, it may not be improper for us to advert to two evils which result from this system.

The first evil is that the upper bed of coal sinks from lack of support, cracks, and in so doing emits gas which passes into the workings below, 'by which some of the most serious accidents of fire-damp have been occasioned.' The second evil is that the policy of plundering the richest seam leads to enormous waste, since the upper beds are rendered useless by 'the continued permeation of water and gas through its fissures, whereby property of great value may be irrecoverably lost to the country'.

Here we have the difference between the narrow technical difficulty (gas in the goaf, damaged safety lamps) and the broad social context (plundering the thickest seam and generating gas as a result) which renders all such technical solutions useless. Faraday did not feel at liberty to deal with the second problem, and it was certainly less easily treated than a specific technical one; but it was the political economy of mining that rendered his own investigations irrelevant. By confining itself to these so-called scientific solutions, the State managed to avoid the type of confrontation that would have ensued if it attempted to legislate on mining policy; yet there was clearly no solution without a confrontation.

The Special Committee of the United Committee of the Coal Trade published its reply to this on 7 February 1845. With the exception of the one real issue Faraday raised—the policy of mining itself—the Faraday/Lyell report appears quite foolish in the light of this reply. The owners addressed themselves to the most sensitive issue first, and in this they were not very successful. They argued that in fact it was quite usual to work the upper seams first (contrary to what Faraday stated), and that 'a capitalist will gladly avail himself of the privilege of winning, at a less amount of outlay, an upper, and therefore more accessible seam of coal, provided that by so doing he can compete successfully'. But after this, their reply degenerated into double-talk. In cases where the under seams are worked first, they continued, upper seams are injured very slightly. When subsidence of a bed takes place, since it is entire, one part of it (they claimed) supports another part, so that it does not disintegrate. As simple observation of a coal bed was enough to refute this, one wonders what the owners thought they were doing. By pointing to mining policy itself, Faraday had unwittingly struck at the central problem germane to accidents.

The major part of their reply, however, constituted an abrupt *volta face* from the position taken by the Haswell owners and their representatives at the inquest. For as a result of Faraday's report, the government had become very enthusiastic about its principal recommendation (withdrawing the gas by means of pipes), which would have been enormously expensive to carry out. What the owners now demonstrated, without much difficulty, was that Faraday really did not understand the nature of a goaf. As the props are withdrawn and the roof sinks, the crush either renders the new ceiling solid or divides it into isolated compartments. There was, then, usually no communication between the goaves, so that the gas was unable to filter through. Gas was thus not necessarily found at the upper reaches, but often in the deeper recesses, and thus Faraday's plan for draining the upper portions 'could not be depended upon as draining it throughout its entire space . . .'. One would have to have a separate apparatus for each goaf, and the estimate for Haswell alone came to twelve miles of pipes at a cost of at least £21,000. The Committee also reported that in addition to Haswell, during the last fourteen years there had been eleven great explosions in the Northumberland and Durham collieries, and in at least ten cases, possibly all eleven, the explosions occurred where the mines were being worked 'in the whole', i.e. the coal face had not been broken into so deeply yet as to create goaves. In short, in these cases goaves had no relation to accidents. Although the owners' discussion of roof-collapse, and the statistics on the explosions of the last fourteen years, were probably correct, one wonders how these matters never came to be mentioned at the inquest itself. 'Scientific' testimony was proving to be remarkably plastic.

In conclusion, the owners stated that although they were sure that the mining schools Faraday advocated would be very desirable, they could not but observe that France and Germany had many such schools, and that 'the mining practice of those countries is in no respect superior to that of Great Britain'. The owners ended their reply on a note of heavy irony, that they were deeply indebted to Lyell and Faraday, and trusted that when the 'intricate practical difficulties [of mining] come to be fully appreciated by them, their eminent acquirements may prove highly beneficial'. Faraday's and Lyell's reply, given in a letter to S. M. Phillipps at the Home Office (printed in the 1845 report) of 27 March 1845 was brief, and unable to refute the major points of the owners' reply to their report.

As an interesting postscript, Matthias Dunn, the viewer excluded by the owners from examining the mine, published a review of the Faraday report, also in 1845.[117] His criticisms are similar to those of the owners, pointing out that the goaf-draining system may be theoretically accurate but is practically useless: faults, varying water levels, and other problems make the pipe system unworkable. But his major criticism of the report focuses on its political naïveté, and it becomes clear why the owners had personal objections to his

[117] Matthias Dunn, *Review of the report of Messrs. Lyell and Faraday, upon the subject of the explosions in coal mines, arising from the catastrophe at Haswell, in September, 1844* (London: Simpkin and Marshall, 1845).

participation at the inquest. Faraday and Lyell, says Dunn, proceed as though 'there is no such thing as prejudice amongst the parties . . .. Are none of them activated by mistaken economy, even to the sacrifice of life and property?'[118] 'I must stop a moment,' he writes,

> to remind these learned gentlemen that profit is the main object of working a coal mine; the lessees, therefore, must and will judge of the capabilities of each respective seam, and the owner of the mine must condescend to grant such terms as the adventurer shall consent to, regardless of any theory, or even the dread of fire-damp to be encountered by such system.[119]

But neither Faraday, nor the Utilitarians, admitted the existence of conflicts of principle.

Of course, Dunn's review went unnoticed, whereas Faraday's report, though never implemented, was acclaimed. For Dunn had pointed out the real problem, Faraday the anodyne. Dunn's criticism revolved around the inadequacy of a system geared to production and profit to deal with human or ecological concerns; Faraday's recommendations effectively said there was no incompatibility between them. It appears, then, that he and Lyell had become fixated on an issue that was actually trivial, and made it the pivot of their entire report, as a result of the testimony of two outside viewers and Lyell's friend Scutchbury, the surveyor. The goaf was not the crucial factor, and their method of remedying the problem was unworkable even if it had been. Nevertheless, the publicity was already achieved, and the impression was that of science and scientific commissions triumphing over social evils.

Referring to the nineteenth century 'revolution in government,' Roy MacLeod writes:

> [I]t is in the applications of science to the purposes of the State . . . that we find the most penetrating examples of technical regulation and control, coordination of national interest and public welfare, and incipient tendencies towards the systematic use of administrative expertise that has become so characteristic of contemporary society.[120]

Certainly, the use of science to effect an increasing rationalization of modern life is a thesis with which we must agree, given the examples of this and the previous chapter; but whether this represents a 'coordination of national interest and public welfare' remains a dubious argument. The reification of modern life by means of a redefined scientific method was motivated by the desire to smooth over structural contradictions, and it was just this process that made the continuation of those contradictions possible. It is not that a conspiracy was at work: there is nothing to suggest that a select number of professionals mapped out a plan to soften the harsher aspects of capitalism. But

[118] ibid., p. 12.
[119] ibid., p. 4.
[120] Roy M. MacLeod, 'The Alkali Acts Administration, 1863–84: the Emergence of the Civil Scientist', *Victorian Studies*, IX (1965), 85.

186 Social Change and Scientific Organization

neither was this accidental. The social problems generated by rapid industrialization created opportunities for the emergence of a priesthood, a large and amorphous number of men who saw positions of influence opening up and needed an ideological justification for the exercise of power. Nothing but science, defined in the way Bentham had already done, could have provided such a justification. Thus was born the dominant psychology of industrial society, the pseudo-science of human problems that would reveal its own intellectual poverty if it ever actually managed to confront causes instead of symptoms. The Royal Institution did not accomplish this on its own, of course, and in fact, after mid-century the whole Utilitarian movement was in abeyance for a time; but the RI did serve as the opening wedge in the *nuova scienza* —what F. R. Leavis has called 'technologico-Benthamite civilization'[121] —and its history reveals the mechanics of how this change came about.

[121] Quoted by Piers Brandon in the *New Statesman*, 2 February 1973.

Conclusion: The Epistemology of Modernization

During the second half of the nineteenth century, the Royal Institution was eclipsed in social importance by the development of scientific curricula at universities and the establishment of laboratories such as the Cavendish. Those things that had made the RI so innovative and vital were being taken over by other institutions. By 1860 there were Owens College, Manchester, the Royal College of Chemistry, and the Royal School of Mines from which to draw experts and consultants. There were medical schools at hospitals, and Brande's 9 a.m. lectures of four decades drew to a close with his resignation in 1852, and were not replaced. The Industrial Revolution had become manageable, or at least manageable enough: it was not in need of constant explanation such as the RI was able to provide. The Utilitarian aspects of science began to draw thinner audiences by 1859, and the concerned Managers had to add a whole host of non-scientific subjects. More and more, the RI, stripped of its once dynamic roles, lapsed back into the amateur tradition. It became increasingly genteel, making no active contribution to England's further development, not even during the First World War, for example. It became, in effect, something of a dodo bird.[1]

There were other factors that pushed it in this direction as well. By 1850 Utilitarianism was on the run: 'Englishmen were in full revolt against the principle of centralization'.[2] The Poor Law Commission was replaced by a more timid Poor Law Board in 1847; Chadwick was fired by the government in 1854; and Lord Shaftesbury, a former advocate of centralization, spoke against it before the Social Science Association in 1858, the same year that the Board of Health came to an end.[3] The professions were also having their troubles. One contemporary relates that Parliamentary practice and committee work, which had attracted so many men to the Bar, had fallen so low in status that the law was now 'a sinking profession'.[4] Just as an agricultural depression had once rendered the improving landlords superfluous, a professional and

[1] *See* the introductions to Vols. X, XI, and XII of the facsimile edition of the Managers' Minutes, by Sophie Forgan, ed. Frank Greenaway (Menston, Yorks.: The Scolar Press, 1976).
[2] Robert M. Gutchen, 'Local Improvements and Centralization in Nineteenth-Century England', *The Historical Journal*, IV (1961), 85.
[3] ibid.
[4] H. Byerley Thomson, *The Choice of a Profession* (London: Chapman and Hall, 1857), p. 29.

Utilitarian 'depression', if it can be so called, struck at the base of much of the RI's *raison d'être*.

In the absence of any coherent programme, the RI naturally fell back on the amateur tradition, although under its new Secretary (as of 1860), Henry Bence Jones, there was an attempt to emphasize the pure research aspect of this.[5] Both he and Faraday had been conducting an unsuccessful campaign in this direction during the 1850s, which received its finest expression in 1870–1 with the publication of two histories that became standard works. In *The Royal Institution: Its Founders and Its First Professors*, Bence Jones created an RI that never existed; in *The Life and Letters of Faraday* he painted a portrait of Faraday which was lopsided in the extreme. This myth-making was not successful in turning the RI into a centre primarily devoted to pure scientific research; but although the effort failed, the myths have remained, and with them a popular conception of science that, it seems to me, is no longer tenable.

The Institution's role as the vehicle of scientific ideology has never been noticed. This in itself is no surprise: science is so much a part of modern industrial society that to talk of it as an ideology, or as the mode of cognition of such society,[6] is akin to talking heresy. For this would imply that our perceptions of reality are relative, rather than uniquely in touch with the truth; that in a social system not defined so completely as ours is by production, 'science would arrive at essentially different concepts of nature and establish essentially different facts'.[7] We have seen how, at the Royal Institution, the vested interests of certain very influential groups led them to emphasize ideologies of science congruent with increased economic production or the control of some of its social consequences, when science was not able to make any real difference for these purposes. Today, it 'naturally' functions in these ways, and the distortion which had been necessary to create this state of affairs is completely forgotten. For example, John Millington both built a treadmill for the Bedfordshire House of Correction and lectured at the RI on prison machinery; Brande was commissioned to calculate Poor Law diets.[8] Yet there was no scientific theory involved in treadmill construction, and the science of nutrition was, in the 1840s, non-existent. The Haswell investigation is probably the most dramatic illustration of this sort of scientific mystification. What is perhaps most frightening about it was that the officials involved (not to say Faraday himself) did not regard the exclusion of dissenting testimony, or the almost immediate exoneration of the mine owners, as a whitewash, because the particular definition of science being used blocked the possibility of seeing the

[5] On this *see* Sophie Forgan, introductions to Vols. XI and XII of the Mangers' Minutes, ed. Frank Greenaway.

[6] Ernest Gellner, *Thought and Change* (Chicago: University Press, 1964), p. 72; quoted in Arnold Thackray, 'Natural Knowledge in Cultural Context: The Manchester Model', *The American Historical Review*, LXXIX (1974), 672.

[7] Herbert Marcuse, *One-Dimensional Man: Studies in the Ideology of Advanced Industrial Society* (Boston: Beacon Press, 1966), p. 167.

[8] J. C. Hippisley, *Prison Labour, &c.* (London: William Nicol, 1823), p. 9; RI Archives, Box 9 (old numbering system), page at end of Brande notebook, 'Table of Composition of Food to be used in Calculating the Poor Law Dietaries', dated December 1849.

colliery disaster as a political problem with a political solution. The scientific method was hardly born in the nineteenth century, of course, but it was during the Industrial Revolution, and through an institution like the RI, that legitimacy was conferred on the scientific management of social problems as a reasonable way of life. In many ways, the legal ideology of science marks the final stage of the modernizing process, because it bestows the aura of objectivity on a very biased status quo. This enormous 'flexibility' of industrial society, the ability to integrate or mitigate all serious opposition, and this scientific 'treatment' of symptoms of a profound malaise which is conveniently ignored—in short, the rapid acceleration of a way of life that can deliver joy to only a fraction of its members—has its origins in the blind scientism of the nineteenth century.

It is, of course, easy to see scientific expertise as necessary, given the specialization required, for the smooth functioning of industrial society. What this overlooks is that experts demand specialization, not the reverse. What must be explored, as István Mészáros notes, is 'the underlying system that gives rise to such "specializations"', not how 'fortunate' society was to somehow have had such men around at this crucial juncture.[9] At least at the Royal Institution, the emergence of certain subgroups (agricultural improvers, Utilitarians) within important classes constitutes the key to permeation of science throughout English life. 'Whatever else ideologies may be,' writes Clifford Geertz, 'they are, most distinctively, maps of problematic social reality and matrices for the creation of collective conscience.'[10] Surely this definition fits the various conceptions of science at the RI perfectly. Thus Monica Grobel writes that to Utilitarians and Whigs, diffusion of scientific knowledge was vital because it was seen as 'the most effective method of reconciling human beings with the facts of the industrial revolution, the lubricating oil in the vast mechanization of life'.[11] The question facing Western society today, or so I would think, is whether we wish to be so reconciled any longer, and whether the dominant culture can ever break free from this not-so-objective definition of scientific objectivity. The perception of reality in scientific terms is not the result of the successful, inevitable progress of the history of ideas. It is, rather, rooted in class society, and in this sense it was the Industrial Revolution that put the Scientific Revolution on the map. Defined as a commodity or as the crux of professional expertise, science was, in the nineteenth century, recreated in capitalism's image. The very success of the latter obscured the ideological roots of science, as might be expected; and the rather obvious decay of industrial society (whether socialist or capitalist) has led us, just as inevitably, to seek to uncover them. This may, indeed, strike at the very foundations of rational knowledge, but if so, I would pose the distinction made in the In-

---

[9] István Mészáros, 'Ideology and Social Science', *The Socialist Register*, 1972, pp. 35–81.
[10] Clifford Geertz, 'Ideology as a Cultural System', in David E. Apter (ed.), *Ideology and Discontent* (New York: The Free Press, 1964), p. 64.
[11] Monica C. Grobel, *The Society for the Diffusion of Useful Knowledge 1826–1846* (Ph.D. Dissertation, University of London, 1932), pp. 2–3.

troduction: rational, or *zweckrational*? The former we cannot live without; the latter is crushing us beneath the weight of its instrumentality. At least on this point, the historian and the prophet might concur: there are new worlds waiting to be born.

Bibliography

Unpublished Documents, Manuscripts, and Archival Materials

1 *At the Royal Institution*
Besides the *MM* and *VM*, there is a bound volume of Rumford MSS and a very large collection of Davy and Faraday MSS, as well as the autobiography of Thomas Webster. Other materials quoted or referred to in the text, such as Bence Jones' copy of the Prospectus of 1800, the letters between Brande and Ure, the Diary of Margery Reid, etc., are kept in the Strong Room.

2 *At the India Office Library and Records*
Bengal Public Consultations
Bombay Commercial Proceedings
Correspondence with India, Despatches to Bombay, Madras, Bengal; Letters Received from Bengal, Madras
Home Correspondence, Miscellaneous Letters Sent
MSS. Eur. E. 11, James Kerr, 'Observations Upon Natural History'
Minutes of the Court of Directors

3 *At the British Museum*
Additional MSS

4 *At the Royal Agricultural Society of England*
Materials preserved from the Board of Agriculture:
Board Minute Books
Rough Minute Books
Registry of Official, Ordinary, Honorary, and Corresponding Members of the Board of Agriculture

5 *At the Library of Wellcome Institute for the History of Medicine*
Faraday correspondence, File 69293

6 *At the Institution of Electrical Engineers*
Faraday Papers

Faraday's *Commonplace Book*
Faraday's MS notes of the Haswell investigation, and correspondence dealing with the incident

7 *At the Royal College of Physicians, Wellcome Library*
Correspondence between P. M. Roget and Alexander Marcet

8 *At University College, London (D. M. S. Watson Library)*
College Correspondence
Chadwick Papers
Society for the Diffusion of Useful Knowledge Letters
Society for the Diffusion of Useful Knowledge, Minutes of Sub-Committees
Society for the Diffusion of Useful Knowledge, Goldsmid Correspondence (2 vols., 1828–34)

9 *At the Library of Trinity College, Cambridge*
Dawson Turner Collection, correspondence with Edward Magrath

10 *At the Royal Society*
Domestic MSS
Miscellaneous MSS
The Diary of Sir Charles Blagden, eight MS volumes
Thomas Bowdler, 'An Account of an Iron Bridge lately erected near Sunderland, by Rowland Burdon M.P. In a Letter from Tho. Bowdler Esq. to Sir Jos. Banks P.R.S', *Letters and Papers*, Decade XI, No. 55.
Faraday's glass notebook
Faraday-Lubbock correspondence, LUB. F.3-F.6

11 *At the Muniment Room, Althorp, Northampton*
Letter from Sir Joseph Banks to the second Earl Spencer, 6 May 1799

12 *At the Principal Probate Registry, Public Record Office*
Will of Edward Goat(e), 4 August 1802

13 *At the John Rylands Library of the University of Manchester*
The Earl Spencer's copy of the first RI Prospectus, Press Mark 13711.6: *Sketch of a Prospectus of the Royal Institution of Great Britain, incorporated by charter, M,DCC,XCIX*

14 *At the Birmingham Public Libraries*
Matthew Boulton Papers, cited as Tew MSS in the text (on loan from the Matthew Boulton Trust)
Boulton and Watt Collection

15 *At the Guildhall Library*
   Court Minute Book, Apothecaries Society, MS. 8200/10
   LI papers, MSS. 2752, 2754, 2758, 3080

16 *Dissertations at the University of London*
   Monica C. Grobel, *The Society for the Diffusion of Useful Knowledge 1826–1846* (Ph.D., 1932)
   Winifred Harrison, *The Board of Agriculture, 1793–1822, with Special Reference to Sir John Sinclair* (M.A., 1955)
   S. G. Smith, *The Contributions to Science of John Bostock, M.D., F.R.S., 1773–1846* (M.Sc., 1953)
   Aubrey A. Tulley, *The Chemical Studies of William Thomas Brande 1788–1866* (M.Sc., 1971)
   Claudio Veliz, *Arthur Young and the English Landed Interest 1784–1814* (Ph.D., 1959)

17 *Parliamentary Papers*
   Cited in the text under the title of the report

Published Books and Articles

A.Z. 'Obstacles to Agricultural Improvement', *The Farmer's Magazine*, III (1802), 305–6.

Abel-Smith, Brian, and Stevens, Robert. *Lawyers and the Courts*. London: Heinemann Educational, 1967.

Abrams, Philip. *The Origins of British Sociology: 1834–1914*. Chicago: University Press, 1968.

Agassi, Joseph. *Faraday as a Natural Philosopher*. Chicago: University Press, 1971.

Allen, William. *Life of William Allen, with selections from his correspondence*. 3 vols. London: Charles Gilpin, 1846.

Althusser, Louis. *For Marx*. Trans. Ben Brewster. New York: Vintage Books, 1970.

Anderson, James. 'To the Right Honourable Lord Sheffield', *Recreations in Agriculture, Natural-History, Arts, and Miscellaneous Literature*, VI (1802), 35–60.

Anderson, R. E. 'Leonard Horner', *DNB*, XXVII (1891), 371–2.

Annals of the Royal Statistical Society 1834–1934. London: The Royal Statistical Society, 1934.

Archer, Mildred. *British Drawings in the India Office Library*. 2 vols. London: H.M.S.O., 1970.

——— 'India and Natural History: the Role of the East India Company 1785–1858', *History Today*, IX (1959), 736–43.

——— *Natural History Drawings in the India Office Library*. London: H.M.S.O., 1962.

Ashton, T. S. *Economic and Social Investigations in Manchester, 1833–1933.* London: P. S. King and Son, 1934.

——— *An Economic History of England: The Eighteenth Century.* London: Methuen, 1955; reprinted, with minor corrections, 1961.

——— *The Industrial Revolution 1760–1830.* New York: Oxford University Press, 1964.

Aubert, Vilhelm (ed.). *Sociology of Law.* Harmondsworth: Penguin Books, 1969.

Babbage, Charles. *The Exposition of 1851.* London: John Murray, 1851.

——— *Reflections Upon the Decline of Science in England, and on some of its causes.* London: B. Fellowes, 1830.

Baker, J. Bernard (ed.). *Pleasure and Pain (1780–1818).* London: John Murray, 1930.

Barnes, Barry. *Scientific Knowledge and Sociological Theory.* London: Routledge and Kegan Paul, 1974.

——— (ed). *Sociology of Science.* Harmondsworth: Penguin Books, 1972.

Barratt, C. R. B. *The History of the Society of Apothecaries.* London: Elliot Stock, 1905.

Barthes, Roland. *Mythologies.* Trans. Annette Lavers. New York: Hill and Wang, 1972.

Barton, D. B. *A History of Copper Mining in Cornwall and Devon.* Truro: Truro Bookshop, 1961.

Beddoes, Thomas (ed.). *Contributions to Physical and Medical Knowledge, principally from the West of England.* Bristol: Biggs and Cottle, 1799.

Bell, Jacob. *A Concise Historical Sketch of the Progress of Pharmacy in Great Britain.* London: John Churchill, 1843.

Bellasis, Edward. *Memorials of Mr. Serjeant Bellasis.* London: Burns and Oates, 1893.

Bellot, H. Hale. *University College London 1826–1926.* London University Press, 1929.

Bellot, H. H. L. 'The Exclusion of Attorneys from the Inns of Court', *The Law Quarterly Review,* XXVI (1910), 137–45.

Berger, Peter L., and Luckmann, Thomas. *The Social Construction of Reality.* Garden City: Doubleday, 1966.

Berman, Morris. 'The Early Years of the Royal Institution 1799–1810: A Re-evaluation', *Science Studies,* II (1972), 205–40.

——— '"Hegemony" and the Amateur Tradition in British Science', *Journal of Social History,* VIII (Winter 1975), 30–50.

Bernard, Thomas. *A Letter to the Honourable and Right Reverend the Lord Bishop of Durham, President of the Society for Bettering the Condition of the Poor, on the Principle and Detail of the Measures Now under the Consideration of Parliament, for Promoting and Encouraging Industry, and for the Relief and Regulation of the Poor.* 2nd ed. London: J. Hatchard, 1807.

Bettany, G. T. 'Augustus Bozzi Granville', *DNB,* XXII (1890), 412–14.

Bevan, G. Phillips. *The London Water Supply*. London: Edward Stanford, 1884.

Biggin, George. 'Experiments to determine the Quantity of tanning Principle and Gallic Acid, contained in the Bark of various Trees', *Phil. Trans.*, LXXXIX (1799), 259–64.

The Biographical Dictionary of the Society for the Diffusion of Useful Knowledge. 4 vols. London: Longman, Brown, Green, and Longman, 1842.

Blaug, Mark. 'The Myth of the Old Poor Law and the Making of the New', *The Journal of Economic History*, XXIII (1963), 151–84.

———— 'The Poor Law Report Reexamined', *The Journal of Economic History*, XXIV (1964), 229–45.

Bowden, Witt. *Industrial Society in England towards the End of the Eighteenth Century*. 2nd ed. London: Frank Cass, 1965.

Bradley, Duane. *Count Rumford*. Princeton: D. Van Nostrand, 1967.

Brande, W. T. 'On the composition and analysis of the inflammable gaseous compounds resulting from the destructive distillation of coal and oil, with some remarks on their relative heating and illuminating powers', *Phil. Trans.*, CX (1820), 11–28.

———— *QJS* (*see* list of abbreviations).

———— 'Resignation of Professor Brande', *Notices of the Proceedings at the Meetings of the Members of the Royal Institution*, I (1852), 168–9.

Braudel, Fernand. *Capitalism and Material Life*. Trans. Miriam Kochan. London: Weidenfeld and Nicolson, 1973.

Brayley, E. W., Jr. *The Utility of the Knowledge of Nature Considered*. London: Baldwin and Cradock, 1831.

Brebner, J. B. '*Laissez-faire* and State Intervention in Nineteenth-century Britain', in E. M. Carus-Wilson (ed.), *Essays in Economic History*, III (1962), 252–62.

Briggs, Asa. *The Making of Modern England*. New York: Harper and Row, 1965.

———— ' "Middlemarch" and the Doctors,' *The Cambridge Journal*, I (1948), 749–62.

Briscoe, John I. *A letter on the nature and effects of the tread-wheel*. London: John Hatchard and Son, 1824.

The British Magazine, Vols. I–II (1800).

Brodie, B. C. (ed.) *Autobiography of the Late Sir Benjamin C. Brodie, Bart.* London: Longman, Green, Longman, Roberts, and Green, 1865.

Brook, Charles. *Thomas Wakley*. London: The Socialist Medical Association, 1962.

Brown, A. F. J. *Essex at Work 1700–1815*. Chelmsford: Tindal Press, 1969.

Brown, Ford K. *Fathers of the Victorians: The Age of Wilberforce*. Cambridge University Press, 1961.

Brown, N. O. *Life Against Death*. Middletown, Conn.: Wesleyan University Press, 1959.

Brown, Sanborn C. (ed.). *Benjamin Thompson—Count Rumford: Count Rum-*

ford on the Nature of Heat. Oxford: Pergamon Press, 1967.

——— *Count Rumford: Physicist Extraordinary.* Garden City: Doubleday, 1962.

Browne, C. A. 'The Life and Chemical Services of Frederick Accum', *Journal of Chemical Education,* II (1925), 829–51, 1008–34, 1140–49.

——— 'Recently Acquired Information Concerning Frederick Accum, 1769–1838', *Chymia,* I (1948), 1–10.

Brundage, Anthony. 'The Landed Interest and the New Poor Law: a reappraisal of the revolution in government', *English Historical Review,* LXXXVII (1972), 27–48.

Burn, W. L. *The Age of Equipoise.* London: George Allen and Unwin, 1969.

Burns, J. H. *Jeremy Bentham and University College.* London: The Athlone Press, 1962.

Burridge, John. *The tanner's key to a new system of tanning sole leather, or the right use of oak bark, &c. stating several most important discoveries and documents, in addition to Sir Humphrey [sic] Davy's experiments and the late Sir Joseph Banks' reports to the Honourable East India Company.* London: Charles Frederick Cock, 1824.

Cameron, H. C. *Sir Joseph Banks.* London: Angus and Robertson, 1952.

Cameron, Rondo. *France and the Economic Development of Europe 1800–1914.* 2nd ed., rev. and abgd. Chicago: Rand McNally, 1966.

Cardwell, D. S. L. *The Organisation of Science in England.* Melbourne: Heinemann, 1957.

[Carlyle, Thomas]. 'Signs of the Times', *The Edinburgh Review,* XLIX (1829), 439–59.

Caroe, A. D. R. *The House of the Royal Institution.* London: The Royal Institution, 1963.

Carr-Saunders, A. M., and Wilson, P. A. *The Professions.* Oxford: Clarendon Press, 1933.

Carter, H. B. *His Majesty's Spanish Flock.* Sydney: Angus and Robertson, 1964.

Cartwright, F. *The English Pioneers of Anaesthesia: Beddoes, Davy, and Hickman.* Bristol: John Wright and Sons, 1952.

Cazamian, Louis. *The Social Novel in England 1830–1850.* Trans. Martin Fido. London: Routledge and Kegan Paul, 1973 reprint.

Chambers, J. D., and Mingay, G. E. *The Agricultural Revolution 1750–1880.* London: B. T. Batsford, 1966.

Chandler, Dean, and Lacey, A. D. *The Rise of the Gas Industry in Britain.* London: British Gas Council, 1969.

Chandra, Prakash. 'The Establishment of the Fort William College', *The Calcutta Review,* LI, Third Series (1934), 160–71.

The Charter, the Act of Parliament and the By-Laws, of the London Institution. London: Waterlow and Sons, 1902.

The Charter and Bye-Laws of the Royal Institution of Great Britain. London:

The Royal Institution, 1803.

The Charter and Bye-Laws of the Royal Institution of Great Britain. London: Savage and Easingwood, 1806.

Checkland, S. G. 'The Prescriptions of the Classical Economists', *Economica*, n.s., XX (1953), 61–72.

Clark, G. Kitson. *The English Inheritance.* London: SCM Press, 1950.

Clarke, Ernest. 'The Board of Agriculture, 1793–1822', *The Journal of the RASE*, Third Series, IX (1898), 1–41.

——— 'John, Fifteenth Lord Somerville', *The Journal of the RASE*, Third Series, VIII (1897), 1–20.

Clarke, George. *A History of the Royal College of Physicians.* 2 vols. Oxford: Clarendon Press, 1966.

Clarke, J. F. *Autobiographical Recollections of the Medical Profession.* London: J. and A. Churchill, 1874.

Clarkson, L. A. 'Leather Crafts in Tudor and Stuart England', *The Agricultural History Review*, XIV (1966), 25–39.

——— 'The Organization of the English Leather Industry in the Late Sixteenth and Seventeenth Centuries', *The Economic History Review*, Series Two, XIII (1960), 245–56.

Clegg, Samuel, Jr. *A Practical Treatise on the Manufacture and Distribution of Coal-Gas.* London: John Weale, 1841.

Clokie, H. D., and Robinson, J. W. *Royal Commissions of Inquiry.* Stanford: University Press, 1937.

Clow, A., and N. L. *The Chemical Revolution.* London: The Batchworth Press, 1952.

Coates, W. H. 'Benthamism, Laissez-Faire, and Collectivism', *The Journal of the History of Ideas*, XI (1950), 357–63.

Coats, W. H. (ed.). *The Classical Economists and Economic Policy.* London: Methuen, 1971.

Cockburn, Lord. *Memorials of His Time.* Edinburgh: Robert Grant and Son, 1946 reprint.

Cohen, E. W. *The Growth of the British Civil Service 1780–1939.* London: Allen and Unwin, 1941.

Cole, G. D. H., and Postgate, Raymond. *The Common People 1746–1946.* 4th ed. London: Methuen, 1968.

Cole, R. J. 'Friedrich Accum (1769–1838). A Biographical Study', *Annals of Science*, VII (1951), 128–43.

Coley, N. G. 'Alexander Marcet (1770–1822), Physician and Animal Chemist', *Medical History*, XII (1968), 394–402.

——— 'The Animal Chemistry Club; Assistant Society to the Royal Society', *Notes and Records of the Royal Society of London*, XXII (1967), 173–85.

Collins, Kins. 'Marx on the English Agricultural Revolution: Theory and Evidence', *History and Theory*, VI (1967), 351–81.

The Commercial and Agricultural Magazine, II (1800).

Communications to the Board of Agriculture, on subjects relative to the husbandry and internal improvement of the country. 7 vols. London, 1797–1813.

Connell, Brian. *Portrait of a Whig Peer*. London: Andre Deutsch, 1957.

Cooper, Thomas. 'William Farish', *DNB*, XVIII (1889), 208.

Copeman, W. S. C. *The Worshipful Society of Apothecaries of London 1617–1967*. London: Pergamon Press, 1967.

Craig, John. *The Mint*. Cambridge: University Press, 1953.

Cranmer-Byng, J. L. (ed.). *An Embassy to China: Being the journal kept by Lord Macartney during his embassy to the Emperor Ch'ien-lung 1793–1794*. London: Longmans, 1962.

Cromwell, Valerie. 'Interpretations of the Nineteenth-Century Administration: An Analysis', *Victorian Studies*, IX (1966), 245–55.

Crowther, J. G. *British Scientists of the Nineteenth Century*. London: Routledge and Kegan Paul, 1935.

Danvers, F. C., *et al. Memorials of Old Haileybury College*. London: Archibald Constable and Company, 1894.

Darvall, Frank O. *Popular Disturbances and Public Order in Regency England*. London: Oxford University Press, 1934.

Davy, Humphry. 'An Account of some Galvanic Combinations, formed by the arrangement of single metallic Plates and Fluids, analogous to the new Galvanic Apparatus of Mr. Volta', *Phil. Trans.*, XCI (1801), 397–402.

———— *The Collected Works of Sir Humphry Davy, Bart*. Ed. John Davy. 9 vols. London: Smith, Elder, 1839.

———— *Elements of Agricultural Chemistry, in a Course of Lectures for the Board of Agriculture*. London: W. Bulmer, 1813.

Reviews of this work: *The Edinburgh Review*, XXII (1814), 251–81; *Gentleman's Magazine*, LXXXIV (1814), Part 1, pp. 466–8; *Monthly Review*, LXXV (1814), 134–42.

———— *Lecture, on the plan which is proposed to adopt for improving the Royal Institution, and rendering it permanent*. London: William Savage, 1810.

———— *Outlines of a Course of Lectures on Chemical Philosophy*. London: The Royal Institution, 1804.

———— *A Syllabus of a Course of Lectures on Chemistry*. London: The Royal Institution, 1802.

Davy, John. *Fragmentary Remains, Literary and Scientific, of Sir Humphry Davy, Bart*. London: John Churchill, 1858.

———— *Memoirs of the Life of Sir Humphry Davy, Bart*. 2 vols. London: Longman, Rees, Orme, Brown, Green, and Longman, 1836.

Dawson, Warren R. (ed.). *The Banks Letters*. London: The British Museum, 1958.

———— *Supplementary Letters of Sir Joseph Banks*. London: Bulletin of the British Museum (Natural History), Historical Series Vol. 3, No. 2, pp. 41–70, 1962.

Deane, Phyllis. *The First Industrial Revolution*. Cambridge University Press, 1965.

——— and Cole, W. A. *British Economic Growth 1688–1959*. 2nd ed. Cambridge University Press, 1967.

De la Rive, Auguste. 'Michael Faraday, his Life and Works', *Phil. Mag.*, XXIV (1867), 409–37.

De Schweinitz, Karl. *England's Road to Social Security*. Philadelphia: University of Pennsylvania Press, 1943.

Dibdin, Thomas F. *Reminiscences of a Literary Life*. 2 vols. London: John Major, 1836.

Dicey, A. V. *Lectures on the Relation between Law and Public Opinion in England during the Nineteenth Century*. London: Macmillan, 1905.

Dickens, Charles. *The Pickwick Papers*. Harmondsworth: Penguin Books, 1972 reprint.

Dickinson, H. W. *Matthew Boulton*. Cambridge: University Press, 1936.

Dinwiddie, James. *Syllabus of a Course of Lectures on Experimental Philosophy*. London: A. Grant, 1789.

Douglas, Mary. *Purity and Danger*. New York: Praeger, 1966.

Dunn, Matthias. *Review of the report of Messrs. Lyell and Faraday, upon the subject of the explosions in coal mines, arising from the catastrophe at Haswell in September, 1844*. London: Simpkin and Marshall, 1845.

Duveen, Denis I. 'Madame Lavoisier', *Chymia*, IV (1953), 13–29.

Dyos, H. J., and Aldcroft, D. H. *British Transport*. Leicester: University Press, 1969.

Eden, Timothy. *Durham*. 2 vols. London: Robert Hale, 1952.

Eisenstadt, N. 'Studies of Modernization and Sociological Theory', *History and Theory*, XIII (1974), 225–52.

Eliot, George. *Middlemarch*. Harmondsworth: Penguin Books, 1965 reprint.

Elliot, Philip. *The Sociology of the Professions*. London: Macmillan, 1972.

Ellis, George E. *Memoir of Sir Benjamin Thompson, Count Rumford*. Boston: American Academy of Sciences, 1871.

'The English Bar and the Inns of Court', *Quarterly Review*, CXXXVIII (1875), 138–76.

'Espinasse, Margaret. 'The Decline and Fall of Restoration Science', *Past and Present*, No. 14 (1958), 71–89.

Evans, Joan. *A History of the Society of Antiquaries*. Oxford University Press, 1956.

Everard, Stirling. *The History of the Gas Light and Coke Company*. London: Ernest Benn, 1949.

Ewen, A. H. 'The Friend of Mankind: A Portrait of Count Rumford', *Proceedings of the Royal Institution of Great Britain*, XL (1964), 186–200.

'Fac-simile of a Letter from Sir H. Davy to Sir F. Baring', 3 October 1805. In RI Collection of *Miscellaneous Reports, &c.*, Series P, Vol. 2.

Falkus, M. E. 'The British Gas Industry before 1850', *The Economic History Review*, Second Series, XX (1967), 494–508.

Faraday, Michael. *Faraday's Diary*. Ed. Thomas Martin. 8 vols. London: G. Bell and Sons, 1932–6.

———— 'On the Manufacture of Glass for Optical Purposes', *Phil. Trans.*, CXX (1830), 1–57.

———— 'On new compounds of carbon and hydrogen, and on certain other products obtained during the decomposition of oil by heat', *Phil. Trans.*, CXV (1825), 440–66.

———— 'Thoughts on Ray-vibrations', *Phil. Mag.*, XXVIII (1846), 345–50.

Finer, S. E. *The Life and Times of Sir Edwin Chadwick*. London: Methuen, 1952.

Fletcher, H. R. *The Story of the Royal Horticultural Society 1804–1968*. London: Oxford University Press, 1969.

Foote, G. A. 'Science and Its Function in Early Nineteenth Century England', *Osiris*, XI (1954), 438–54.

———— 'Sir Humphry Davy and his Audience at the Royal Institution', *Isis*, XLIII (1952), 6–12.

Foster, Charles I. *An Errand of Mercy*. Chapel Hill: University of North Carolina Press, 1960.

Foucault, Michel. *The Order of Things*. New York: Vintage Books, 1973.

Fox, Hon. Henry Edward. *The Journal of the Hon. Henry Edward Fox*. Ed. the Earl of Ilchester. London: Thornton Butterworth, 1923.

Frazer, R. Watson. *Notes on the History of the London Institution prepared mostly from the minutes*. London: Waterlow and Sons, 1905.

Freund, Julien. *The Sociology of Max Weber*. Trans. Mary Ilford. Harmondsworth: Penguin Books, 1972.

Fullmer, J. Z. 'Davy's Biographers: Notes on Scientific Biography', *Science*, CLV (20 January 1967), 285–91.

———— 'Davy's Sketches of His Contemporaries', *Chymia*, XII (1967), 127–50.

———— 'Humphry Davy's Adversaries', *Chymia*, VIII (1962), 147–64.

———— 'Humphry Davy's Critical Abstracts', *Chymia*, IX (1964), 97–115.

———— *Sir Humphry Davy's Published Works*. Cambridge: Harvard University Press, 1969.

———— Letter to the Editor, *Scientific American*, CCIII (August 1960), 12–14.

Furber, Holden. *John Company at Work*. Cambridge: Harvard University Press, 1948.

Fussell, G. E. 'Science and Practice in Eighteenth-Century British Agriculture', *Agricultural History*, XLIII (1969), 7–18.

Gage, A. T. *A History of the Linnean Society of London*. London: Taylor and Francis, 1938.

Garnett, Richard. 'Thomas Garnett, M.D.', *DNB*, XXI (1890), 7–8.

Garnier, Russell. *History of the English Landed Interest*. 2nd ed. 2 vols.

London: Swan Sonnenschein and Company, 1908.

Gazley, John. *The Life of Arthur Young.* Philadelphia: American Philosophical Society, 1973.

Geertz, Clifford. 'Ideology as a Cultural System', in David E. Apter (ed.), *Ideology and Discontent.* New York: The Free Press, 1964.

Geikie, Archibald. *Life of Sir Roderick Murchison.* 2 vols. London: John Murray, 1875.

Gellner, Ernest. *Thought and Change.* Chicago: University Press, 1964.

George, M. D. *England in Transition.* Harmondsworth: Penguin Books, 1965 reprint.

——— *London Life in the Eighteenth Century.* London: Kegan Paul, Trench, Trubner and Company, 1925.

Gillispie, Charles C. *Genesis and Geology.* New York: Harper and Row, 1959.

——— 'The Natural History of Industry', *Isis*, XLVIII (1957), 398–407.

Gisborne, Thomas. *An Enquiry into the Duties of Men in the Higher and Middle Classes of Society in Great Britain.* 2 vols. London: J. Davis, 1795.

Gladstone, J. H. *Michael Faraday.* London: Macmillan, 1872.

Goffman, Erving. 'Role Distance', in *Encounters.* Harmondsworth: Penguin Books, 1972.

Gordon, Alexander. 'John Glas', *DNB*, XXI (1890), 417–18.

Granville, A. B. *Autobiography of A. B. Granville.* Ed. Paulina B. Granville. 2 vols. London: Henry S. King and Company, 1874.

Gray, B. Kirkman. *A History of English Philanthropy.* London: P. S. King and Son, 1905.

Greig, James (ed.). *The Farington Diary.* 8 vols. London: Hutchinson, 1808–9.

Griffiths, Percival. *The British Impact on India.* London: Macdonald, 1952.

Griggs, Earl L. (ed.). *Collected Letters of Samuel Taylor Coleridge.* 2 vols. Oxford: Clarendon Press, 1956.

Grob, Gerald N. 'The Political System and Social Policy in the Nineteenth Century: Legacy of the Revolution', delivered at the National Archives Conference on Research in the History of the American Revolution, Washington, D.C., 15–16 November 1973.

Gutchen, Robert M. 'Local Improvements and Centralization in Nineteenth-Century England', *The Historical Journal*, IV (1961), 85–96.

Habakkuk, H. J. 'Economic Functions of English Landowners in the Seventeenth and Eighteenth Centuries', *Explorations in Entrepreneurial History*, VI (December 1953), 92–102.

Haber, L. F. *The Chemical Industry During the Nineteenth Century.* Oxford: Clarendon Press, 1958.

Hadfield, Robert A. *Faraday and his Metallurgical Researches.* London: Chapman and Hall, 1931.

——— 'A Research on Faraday's "Steel and Alloys"', *Phil. Trans.*, Series A, CCXXX, 221–92.

Halévy, Elie. *The Growth of Philosophic Radicalism.* Trans. Mary Morris.

London: Faber and Faber, 1972 reprint.

———— *History of the English People in the Nineteenth Century.* Vol. I: *England in 1815.* Trans. E. I. Watkin and D. A. Barker. 2nd ed. London: Ernest Benn, 1961 reprint.

Hamburger, Joseph. *Intellectuals in Politics.* New Haven: Yale University Press, 1965.

Hamilton, Bernice. 'The Medical Profession in the Eighteenth Century', *The Economic History Review*, Second Series, IV (1951), 141–69.

Hammond, J. L., and Barbara. *The Town Labourer.* Garden City: Doubleday, 1968 reprint.

———— *The Village Labourer.* 4th ed. London: Longmans, Green and Company, 1966 reprint.

Handley, Henry. *A Letter to the Earl Spencer . . . on the Formation of a National Agricultural Institution.* 2nd ed. London: James Ridgway and Sons, 1838.

Handley, James E. 'Sir John Sinclair', *The Innes Review*, VIII (1957), 5–18.

Harding, Alan. *A Social History of English Law.* Harmondsworth: Penguin Books, 1966.

Harrison, J. F. C. *The Early Victorians.* London: Cox and Wyman, 1971.

Hart, Jenifer. 'Nineteenth-Century Social Reform: A Tory Interpretation of History', *Past and Present*, No. 31 (July 1965), 39–61.

Hartley, Harold. *Humphry Davy.* London: Thomas Nelson and Sons, 1966.

———— 'Sir Humphry Davy, Bt., P.R.S.', *Proceedings of the Royal Society of London*, Series A, CCLV (1960), 153–80.

Hasbach, W. *A History of the English Agricultural Labourer.* Trans. Ruth Kenyon. London: P. S. King and Son, 1908.

Hatchett, Charles. 'Chemical Experiments on Zoophytes; with some Observations on the Component Parts of Membrane', *Phil. Trans.*, XC (1800), 327–402.

Hayes, R. C. 'Robert R. Livingston', *Dictionary of American Biography*, VI (1933), 320–5.

Hays, J. N. 'Science in the City: the London Institution, 1819–40', *British Journal for the History of Science*, VII (1974), 146–62.

Herapath, William. 'Experiments on Oil and Coal Gas', *Phil. Mag.*, LXI (1823), 424–31.

Hill, Christopher. *Reformation to Industrial Revolution.* Harmondsworth: Penguin Books, 1969.

Hindle, R. S. E. *The British Penal System 1773–1950.* London: Gerald Duckworth and Company, 1951.

Hippisley, J. C. *Prison Labour, &c.* London: William Nicol, 1823.

An Historical Account of the London Institution. London: reprinted from the introductory matter prefixed to the class catalogue, 1835.

Hobsbawm, E. J. *Industry and Empire.* London: Weidenfeld and Nicolson, 1968.

———— and Rudé, George. *Captain Swing.* London: Lawrence and Wishart,

1969.

Hoffman, Walther G. *British Industry 1700–1950*. Trans. W. O. Henderson and W. H. Chaloner. Oxford: Basil Blackwell, 1955.

Holdsworth, William S. *Charles Dickens as a Legal Historian*. New Haven: Yale University Press, 1928.

Holford, George. *An Account of the General Penitentiary at Millbank*. London: C. and J. Rivington, 1828.

Holland, Elizabeth, Lady. *The Journal of Elizabeth Lady Holland*. Ed. the Earl of Ilchester. 2 vols. London: Longmans, Green, 1908.

Holloway, S. W. F. 'The Apothecaries' Act, 1815: A Reinterpretation', *Medical History*, X (1966), 107–29 and 221–36.

——— 'Medical Education in England, 1830–1858: A Sociological Analysis', *History*, XLIX (1964), 299–324.

Holzman, James M. *The Nabobs in England*. New York: 1926.

Horner, Leonard (ed.). *Memoirs and Correspondence of Francis Horner, M.P.* 2 vols. Boston: Little, Brown and Company, 1853.

Houghton, W. E. *The Victorian Frame of Mind*. New Haven: Yale University Press, 1957.

Howard, D. L. *English Prisons*. London: Methuen, 1960.

Howse, Ernest M. *Saints in Politics*. Toronto: University Press, 1952.

Hudson, Kenneth. *Patriotism with Profit: British Agricultural Societies in the Eighteenth and Nineteenth Centuries*. London: Hugh Evelyn, 1972.

Hughes, Edward. 'The Professions in the Eighteenth Century', *Durham University Journal*, n.s., XIII, ii (1952), 46–55.

Hume, L. J. 'Jeremy Bentham and the Nineteenth-Century Revolution in Government', *The Historical Journal*, X (1967), 361–75.

Hunt, Robert. 'William Thomas Brande', *DNB*, VI (1886), 216–18.

Huntington, Samuel P. *Political Order in Changing Societies*. New Haven: Yale University Press, 1968.

James, T. E. 'Rumford and the Royal Institution: A Retrospect', *Nature*, CXXVIII (1931), 476–81.

Janik, Allan, and Toulmin, Stephen. *Wittgenstein's Vienna*. New York: Simon and Schuster, 1973.

Jay, Martin. *The Dialectical Imagination*. Boston: Little, Brown and Company, 1973.

Jeffreys, Alan E. *Michael Faraday: A List of His Lectures and Published Writings*. London: Chapman and Hall, 1960.

Jones, E. L. (ed.). *Agriculture and Economic Growth in England 1650–1815*. London: Methuen, 1967.

Jones, Henry Bence. *An Autobiography*. London: Crusha and Son, 1929.

——— *The Life and Letters of Faraday*. 2 vols. London: Longmans, Green and Company, 1870.

——— *The Royal Institution: Its Founder and Its First Professors*. London: Longmans, Green and Company, 1871.

Journals of the Royal Institution of Great Britain, Vols. I–II, 1802–3.

Judd, Gerrit P. *Members of Parliament 1734–1832*. New Haven: Yale University Press, 1955.

Keeble, Frederick. 'Humphry Davy, Farmer', *Proceedings of the Royal Institution of Great Britain*, XXX (1937–9), 325–38.

Kelly, Thomas. *A History of Adult Education in Great Britain*. Liverpool: University Press, 1962.

Kendall, James. *Humphry Davy: 'Pilot' of Penzance*. London: Faber and Faber, 1954.

Kent, Charles. 'Edward Bellasis', *DNB*, IV (1885), 180–2.

Kerrison, Robert M. *An Inquiry into the Present State of the Medical Profession in England*. London: Longman, Hurst, Rees, Orme, and Brown, 1814.

———— *A Letter to the Rt. Hon. Robert Peel . . . on the Supply of Water to the Metropolis*. London: Thomas Butcher, 1828.

———— *Observations and reflections on the bill now in progress through the House of Commons, for 'better regulating the medical profession as far as regards apothecaries'*. London: Longman, Hurst, Rees, Orme, and Brown, 1815.

King, Charles R. (ed.). *The Life and Correspondence of Rufus King*. 6 vols. New York: G. P. Putnam's Sons. Vol. III: 1896.

Knight, David M. *Atoms and Elements*. London: Hutchinson, 1967.

Lamb, M. C. 'Leather', *Encyclopaedia Britannica*, 14th ed., XII (1968), 845.

The Lancet, issues for 1823–39.

Landes, David. *The Unbound Prometheus*. Cambridge University Press, 1969.

Larwood, H. J. C. 'Science in India before 1850', *British Journal of Educational Studies*, VII (1958), 36–49.

———— 'Western Science in India before 1850', *Journal of the Royal Asiatic Society of Great Britain and Ireland*, 1962, pp. 62–76.

Latham, P. M. *An Account of the Disease Lately Prevalent at the General Penitentiary*. London: Thomas and George Underwood, 1825.

Lazonick, William. 'Karl Marx and Enclosures in England', *The Review of Radical Political Economics*, VI (Summer 1974), 1–59.

[Lehrburger, E.] Larsen, Egon. *An American in Europe*. New York: The Philosophical Library, 1953.

Lelièvre and Pelletier, 'Rapport au Comité de Salut public, sur les nouveaux moyens de tanner les cuirs, proposés par le cit. Armand Seguin', *Annales de Chimie*, XX (1797), 15–18.

Le Marchant, Denis. *Memoir of John Charles Viscount Althorp Third Earl Spencer*. London: Richard Bentley and Son, 1876.

Lennard, Reginald. 'English Agriculture under Charles II: The Evidence of the Royal Society's "Enquiries"', *The Economic History Review*, IV (1932), 23–45.

Levere, Trevor H. 'Faraday, Matter, and Natural Theology—Reflections on an

Unpublished Manuscript', *British Journal for the History of Science*, IV (1968), 95–107.

Lewis, Roy, and Maude, Angus. *Professional People*. London: Phoenix House, 1952.

Lichtheim, George. *From Marx to Hegel*. New York: The Seabury Press, 1971.

Lipschutz, Daniel E. 'The Water Question in London, 1827–1831', *Bulletin of the History of Medicine*, XLII (1968), 510–26.

Loch, James. *Memoir of George Granville, Late Duke of Sutherland, K. G.* London: 1834.

Lomas, Peter. *True and False Experience*. London: Allen Lane, 1973.

Lowe, George. 'Remarks on an Article entitled "A few Facts relating to Gas Illumination", published in No. 14 of *The Quarterly Journal*', *Phil. Mag.*, IV (1820), 37–46.

Lyell, Katherine M. (ed.). *Memoir of Leonard Horner*. 2 vols. London: Women's Printing Society, 1890.

Lyons, Henry. *The Royal Society 1660–1940*. Cambridge University Press, 1944.

MacDonagh, Oliver. 'The Nineteenth-Century Revolution in Government: a Reappraisal', *The Historical Journal*, I (1958), 52–67.

McKendrick, Neil. 'The Role of Science in the Industrial Revolution: a Study of Josiah Wedgwood as Scientist and Industrial Chemist', in M. Teich and R. Young (eds.), *Changing Perspectives in the History of Science*. London: Heinemann Educational, 1973.

MacLeod, R. M. 'The Alkali Acts Administration, 1863–84: the Emergence of the Civil Scientist', *Victorian Studies*, IX (1965), 85–112.

———— 'Government and Resource Conservation: the Salmon Acts Administration, 1860–1886', *Journal of British Studies*, VII (1968), 114–50.

———— 'The Royal Society and the Government Grant: Notes on the Administration of Scientific Research, 1849–1914', *The Historical Journal*, XIV (1971), 323–58.

———— 'Science and Government in Victorian England: Lighthouse Illumination and the Board of Trade, 1866–1886', *Isis*, LX (1969), 5–38.

———— 'Statesmen Undisguised', *The American Historical Review*, LXXVIII (1973), 1386–1405.

Mailly, E. 'The Royal Institution of Great Britain', trans. C. A. Alexander. *Report of the Board of Regents of the Smithsonian Institution* (1867), pp. 203–26.

Mannheim, Karl. *Ideology and Utopia*. Trans. Louis Wirth and Edward Shils. New York: Harcourt, Brace and World, reprint of 1936 edition.

Mantoux, Paul. *The Industrial Revolution in the Eighteenth Century*. Trans. Marjorie Vernon. Rev. ed. New York: Harper and Row, 1961 reprint.

Marcuse, Herbert. *One-Dimensional Man: Studies in the Ideology of Advanced Industrial Society*. Boston: Beacon Press, 1966.

Marsden, William. *A Brief Memoir of the Life and Writings of the Late*

William Marsden. London: J. L. Cox and Sons, 1838.

Martin, Thomas. 'Early Years at the Royal Institution', *British Journal for the History of Science*, II (1964), 99–115.

———— 'Origins of the Royal Institution', *British Journal for the History of Science*, I (1962), 49–64.

———— *The Royal Institution*. 3rd ed. rev. London: The Royal Institution, 1961.

Marx, Karl. *Capital*. Trans. Samuel Moore and Edward Aveling. 3rd ed. 2 vols. London: George Allen and Unwin, 1890.

———— and Engels, Frederick. *The German Ideology*, ed. C. J. Arthur. London: Lawrence and Wishart, 1970.

Mason, Stephen F. *A History of the Sciences*. New rev. ed. New York: Collier Books, 1962.

Mathias, Peter. *The First Industrial Nation*. London: Methuen, 1969.

Matthews, A. W. *A Biography of William Matthews*. London: privately printed, 1899.

Matthews, L. G. *History of Pharmacy in Britain*. London: E. and S. Livingstone, 1962.

Matthews, William. *An Historical Sketch of the Origin, Progress, & Present State of Gas-Lighting*. London: Rowland Hunter, 1827.

———— *A letter to one of the proprietors of the Grand Junction Water Company, on the subject of the Right Hon. Mr. Peel's letter to the water companies of the metropolis*. London: R. Hunter, 1830.

Meacham, Standish. *Henry Thornton of Clapham*. Cambridge: Harvard University Press, 1964.

Medd, Patrick. *Romilly*. London: Collins, 1968.

Merton, Robert K. *Science, Technology and Society in Seventeenth-Century England*. New York: Harper and Row, 1970 reprint.

Merz, John T. *A History of European Thought in the Nineteenth Century*. 4 vols. New York: Dover Publications, 1965 reprint.

Mészáros, István. 'Ideology and Social Science', *The Socialist Register*, 1972, pp. 35–81.

Meteyard, Eliza. *A Group of Englishmen*. London: Longmans, Green and Company, 1971.

'Metropolis Water Supply', *Fraser's Magazine for Town and Country*, X (1834), 561–72.

Midwinter, E. C. *Victorian Social Reform*. London: Longmans, 1968.

Miles, Wyndham D. '"Sir Humphrey Davie, the Prince of Agricultural Chemists"', *Chymia*, VII (1961), 126–34.

Miller, Samuel. *Observations on the necessity for continuous proceedings in The Offices of the Masters in Chancery*. London: S. Sweet, 1848.

Millerson, Geoffrey. *The Qualifying Associations*. London: Routledge and Kegan Paul, 1964.

Minchinton, W. E. 'Agricultural Returns and the Government during the Napoleonic Wars', *The Agricultural History Review*, I (1953), 29–43.

———— 'The Merchants in England in the Eighteenth Century', *Explorations in Entrepreneurial History*, X (1957), 62–71.

Mingay, G. E. 'The "Agricultural Revolution" in English History: A Reconsideration', *Agricultural History*, XXXVII (1963), 123–33.

———— 'Dr. Kerridge's "Agricultural Revolution"': A Comment', *Agricultural History*, XLIII (1969), 477–81.

———— *Enclosure and the Small Farmer in the Age of the Industrial Revolution.* London: Macmillan, 1968.

———— *English Landed Society in the Eighteenth Century.* London: Routledge and Kegan Paul, 1963.

———— 'The Large Estate in Eighteenth-Century England', First International Conference on Economic History, Stockholm, 1960, *Contributions.* Paris: Mouton, 1960, pp. 367–83.

Mitchison, Rosalind. *Agricultural Sir John.* London: Geoffrey Bles, 1962.

———— 'The Old Board of Agriculture (1793–1822)', *English Historical Review*, LXXIV (1959), 41–69.

Montefiore, Claude Joseph Goldsmid. 'Sir Isaac Lyon Goldsmid', *DNB*, XXII (1890), 83–4.

Moore, Barrington, Jr. *Social Origins of Dictatorship and Democracy.* Harmondsworth: Penguin Books, 1966.

Mullett, Charles F. 'Augustus Bozzi Granville: A Medical Knight-Errant', *Journal of the History of Medicine and Allied Sciences*, V (1950), 251–68.

Musson, A. E., and Robinson, Eric. *Science and Technology in the Industrial Revolution.* Manchester: University Press, 1969.

Namier, Lewis B., and Brooke, John (eds.). *The History of Parliament: The House of Commons, 1754–1790.* 3 vols. London: H.M.S.O., 1964.

Neale, R. S. *Class and Ideology in the Nineteenth Century.* London: Routledge and Kegan Paul, 1972.

Newman, Charles. *The Evolution of Medical Education in the Nineteenth Century.* London: Oxford University Press, 1957.

[Nisbet, W.]. *A Picture of the Present State of the Royal College of Physicians of London.* London: Sherwood, Neely, and Jones, 1817.

O'Brien, Edward. *The Lawyer.* London: William Pickering, 1842.

Overton, John H. *The Evangelical Revival in the Eighteenth Century.* New York: Anson D. F. Randolph and Company, 1886.

Owen, David. *English Philanthropy 1660–1960.* London: Oxford University Press, 1965.

Paris, John A. *The Life of Sir Humphry Davy, Bart.* 2 vols. London: Henry Colburn and Richard Bentley, 1831.

Parker, R. A. C. 'Coke of Norfolk and the Agricultural Revolution', *Economic History Review*, Second Series, VIII (1955–6), 156–66.

Parkinson, C. N. *Trade in the Eastern Seas 1793–1813.* Cambridge University Press, 1937.

Parris, Henry. *Constitutional Bureaucracy*. London: George Allen and Unwin, 1969.

———— 'The Nineteenth-Century Revolution in Government: A Reappraisal Reappraised', *The Historical Journal*, III (1960), 17–37.

Parsons, Talcott. *Essays in Sociological Theory*. Rev. ed. Glencoe, Ill.: The Free Press, 1963.

Peacock, George. *Life of Thomas Young*. London: John Murray, 1855.

Perkin, Harold. *The Origins of Modern English Society 1780–1880*. London: Routledge and Kegan Paul, 1969.

Phillips, C. H. *The East India Company 1784–1834*. Manchester: University Press, 1940; reprinted, with minor corrections, 1961.

———— (ed.). *The Correspondence of David Scott, Director and Chairman of the East India Company, Relating to Indian Affairs 1787–1805*. 2 vols. London: Royal Historical Society, 1951.

Plan and Bye-Laws of the London Institution, for the Advancement of Literature and the Diffusion of Useful Knowledge. London: Phillips and Fardon, 1806.

Plumb, J. H. *England in the Eighteenth Century*. Harmondsworth: Penguin Books, 1950.

Pole, William. *Some Short Reminiscences of Events in my Life and Work*. London: privately printed, 1898.

Political Economy Club, Centenary Volume. London: Macmillan, 1921.

Poynter, J. R. *Society and Pauperism*. London: Routledge and Kegan Paul, 1969.

Pritchard, Earl H. 'The Instruction of the East India Company to Lord Macartney on His Embassy to China and His Reports to the Company, 1792–4', *Journal of the Royal Asiatic Society of Great Britain and Ireland*, 1938, pp. 201–30, 375–96, and 493–509.

The Prospectus, Charter, Ordinances, and Bye-Laws, of the Royal Institution of Great Britain. London: W. Bulmer, 1800.

Prospectus, the Society for the Diffusion of Useful Knowledge. London: William Clowes, n.d.

Prothero, Rowland E. (Baron Ernle). *The Pioneers and Progress of English Farming*. London: Longmans, Green and Company, 1888.

Proudfoot, W. J. *Biographical Memoir of James Dinwiddie*. Liverpool: Edward Howell, 1868.

Pryde, George S. *Scotland from 1603 to the present day*. London: Thomas Nelson and Sons, 1962.

Purver, Margery. *The Royal Society: Concept and Creation*. London: Routledge and Kegan Paul, 1967.

Radzinowicz, Leon. *A History of English Criminal Law and its Administration from 1750*. 4 vols. London: Stevens and Sons, 1948–68.

Raistrick, Arthur. *Quakers in Science and Industry*. London: The Bannisdale Press, 1950.

Ranking, G. S. A. 'History of the College of Fort William,' *Bengal Past and Present*, VII (1911), 1–29; XXI (1920), 160–200; XXIV (1922), 112–38.

Raynor, John. *The Middle Class*. London: Longmans, Green and Company, 1969.

Reader, W. J. *Professional Men*. New York: Basic Books, 1966.

The Record of the Royal Society of London. 3rd ed. London: Oxford University Press, 1912.

Redlich, Fritz. 'Science and Charity: Count Rumford and His Followers', *International Review of Social History*, XVI (1971), 184–216.

Regulations to be Observed by Students intending to qualify themselves to practice as Apothecaries, in England and Wales. London: Gilbert and Rivington, 1833.

Reith, Charles. *A New Study of Police History*. London: Oliver and Boyd, 1956.

Renner, Karl. *The Institutions of Private Law and their Social Functions*. Trans. Agnes Schwarzchild. London: Routledge and Kegan Paul, 1949.

Rickman, John (ed.). *Life of Thomas Telford, written by himself*. London: James and Luke G. Hansard and Sons, 1838.

Riley, James F. *The Hammer and the Anvil*. Clapham, via Lancaster, Yorks.: The Dalesman Publishing Company, 1954.

Roberts, David. 'Jeremy Bentham and the Victorian Administrative State', *Victorian Studies*, II (1959), 193–210.

Roberts, William. *Memoirs of the Life and Correspondence of Mrs. Hannah More*. 2 vols. New York: Harper and Brothers, 1835.

Robson, Robert. *The Attorney in Eighteenth-Century England*. Cambridge University Press, 1959.

Rodgers, Betsy. *Cloak of Charity*. London: Methuen, 1949.

Roebuck, Thomas. *The Annals of the College of Fort William*. Calcutta: Philip Pereira, 1819.

Roget, P. M., *et al.* 'Supply of Water in the Metropolis', *London Medical Gazette*, II (1828), 271–5 and 305–8.

Rolt, L. T. C. *Isambard Kingdom Brunel*. Harmondsworth: Penguin, 1970.

Rosen, George. 'Economic and Social Policy in the Development of Public Health', *Journal of the History of Medicine and Allied Sciences*, VIII (1953), 406–30.

———— 'Problems in the Application of Statistical Analysis to Questions of Health: 1700–1880', *Bulletin of the History of Medicine*, XXIX (1955), 27–45.

Roszak, Theodore. *The Making of a Counter Culture*. Garden City: Doubleday, 1969.

Ruggles, Thomas. *The Barrister: or, Strictures on the Education Proper for the Bar*. 2nd ed. London: W. Clarke and Sons, 1818.

Rules, Orders, and Premiums, of the Bath and West of England Society, for the Encouragement of Agriculture, Arts, Manufactures, and Commerce. Bath: R. Cruttwell, 1797.

Rumford, Count. *See* Thompson, Benjamin.

Russell, Colin A. 'The Electrochemical Theory of Sir Humphry Davy', *Annals of Science*, XV (1959), 1–25.

Russell, E. John. *A History of Agricultural Science in Great Britain*. London: George Allen and Unwin, 1966.

Russell, Rollo (ed.). *Early Correspondence of Lord John Russell*. 2 vols. London: T. Fisher Unwin, 1913.

Sandford, Mrs Henry. *Thomas Poole and His Friends*. 2 vols. London: Macmillan, 1888.

Schofield, Robert E. *Lunar Society of Birmingham*. Oxford: Clarendon Press, 1963.

Scott, E. L. 'William Thomas Brande', in C. C. Gillispie (ed.), *Dictionary of Scientific Biography*, II (1970), 420–1.

Senn, P. R. 'The Earliest Use of the Term "Social Science"', *The Journal of the History of Ideas*, XIX (1958), 568–70.

Sennett, Richard, and Cobb, Jonathan. *The Hidden Injuries of Class*. New York: Vintage Books, 1973.

Shapin, S. A., and Thackray, Arnold. 'Prosopography as a Research Tool in History of Science: the British Scientific Community 1700–1900', *History of Science*, XII (1974), 1–28.

Sharlin, Harold I. *The Convergent Century*. London: Abelard-Schuman, 1966.

Sherwin, Oscar. 'An Eighteenth Century Beveridge Planner', *The American Historical Review*, LII (1947), 281–90.

Siegfried, Robert. 'The Chemical Philosophy of Humphry Davy', *Chymia*, V (1959), 193–201.

Sinclair, J. *Memoirs of the Life and Works of the late Right Hon. Sir John Sinclair, Bart.* 2 vols., Edinburgh, 1837.

Sinclair, John. *Account of the Origin of the Board of Agriculture and Its Progress for Three Years after Its Establishment*. London: W. Bulmer, 1796.

———— *Essays on Miscellaneous Subjects*. London: T. Cadell, Jr., and W. Davies, 1802.

———— (ed.). *The Correspondence of the Right Honourable Sir John Sinclair, Bart.* 2 vols. London: Henry Colburn and Richard Bentley, 1831.

Slater, Philip. *Earthwalk*. Garden City: Doubleday, 1974.

Smiles, Samuel. *Lives of Boulton and Watt*. 2nd ed. London: John Murray, 1866.

———— *A Publisher and His Friends: Memoir and Correspondence of the Late John Murray*. 2 vols. London: John Murray, 1891.

Society for the Diffusion of Useful Knowledge. London: George Taylor, 1832.

Somerville, Lord John. *The System Followed During the Last Two Years by the Board of Agriculture*. 2nd ed. London: W. Miller, 1800.

Sparrow, W. J. 'The Break between Count Rumford and the Royal Institution', *Proceedings of the Tenth International Congress of the History of Science*, Vol. I, pp. 307–9. Ithaca, 1962.

———— 'Early Days at the Royal Institution: A Forgotten Experiment in Technological Education', *Educational Review*, VI (1954), 202–7.

———— *Knight of the White Eagle*. London: Hutchinson, 1964.

Spiers, C. H. 'Sir Humphry Davy and the Leather Industry', *Annals of Science*, XXIV (1968), 99–113.

———— 'William Thomas Brande, Leather Expert', *Annals of Science*, XXV (1969), 179–201.

Spring, David. 'Aristocracy, Social Structure, and Religion in the Early Victorian Period', *Victorian Studies*, VI (1963), 263–80.

———— 'The Clapham Sect: Some Social and Political Aspects', *Victorian Studies*, V (1962), 35–48.

Stansky, Peter (ed.). *The Victorian Revolution*. New York: Franklin Watts, 1973.

Statement by the Council of the University of London, explanatory of the Nature and Objects of the Institution. London: Richard Taylor, 1827.

A Statement by the Society of Apothecaries, on the Subject of their Administration of the Apothecaries' Act. London: Samuel Highley, 1844.

Staunton, George L. *An Historical Account of the Embassy to the Emperor of China*. London: John Stockdale, 1797.

[Stephen, George]. *Adventures of an Attorney in Search of a Practice*. London: Saunders and Otley, 1839.

Stephen, Leslie. *The English Utilitarians*. 3 vols. London: Duckworth and Company, 1900.

Stern, Walter M. 'Control *v.* Freedom in Leather Production from the Early Seventeenth to the Early Nineteenth Century', *The Guildhall Miscellany*, II (1968), 438–58.

Stirling, A. M. W. *Coke of Norfolk and His Friends*. 2 vols. London: John Lane The Bodley Head, 1907.

Stone, Lawrence. 'Prosopography', *Daedalus*, C (Winter 1971), 46–79.

'Supply of Water in the Metropolis', *The Lancet* , 21 June 1828, pp. 372–7.

Sutherland, Gillian (ed.). *Studies in the Growth of Nineteenth-Century Government*. London: Routledge and Kegan Paul, 1972.

Sutherland, Lucy. 'The City of London in Eighteenth-Century Politics', in Richard Pares and A. J. P. Taylor, *Essays Presented to Sir Lewis Namier*. London: Macmillan, 1956, pp. 49–74.

———— *The East India Company in Eighteenth Century Politics*. Oxford: Clarendon Press, 1952.

Swainson, William. *A Preliminary Discourse on the Study of Natural History*. London: Longman, Rees, Orme, Brown, Green, and Longman, 1834.

'Tanning', *Encyclopaedia Britannica*, 3rd ed., XVIII, Part 1 (1788–97), 308.

Taylor, A. J. *Laissez-faire and State Intervention in Nineteenth-century Britain*. London: Macmillan, 1972.

Taylor, E. G. R. *The Mathematical Practitioners of Hanoverian England 1714–1840*. Cambridge University Press, 1966.

Taylor, F. S. *A History of Industrial Chemistry.* London: Heinemann, 1957.

Technology and the Frontiers of Knowledge. Garden City: Doubleday, 1973.

Thackray, Arnold. 'Natural Knowledge in Cultural Context: The Manchester Model', *The American Historical Review*, LXXIX (1974), 672–709.

————— and Mendelsohn, Everett (eds.). *Science and Values.* New York: Humanities Press, 1974.

'Thames Water Question', *Westminster Review*, XII (1830), 31–42.

Thompson, Benjamin. *Essays, Political, Economical and Philosophical.* 4 vols. London: T. Cadell, Jr., and W. Davies, 1796–1812.

————— *The Complete Works of Count Rumford.* 4 vols. Boston: American Academy of Arts and Sciences, 1875.

[Thompson, E. P.]. 'Land of Our Fathers', *Times Literary Supplement*, CCCXC (16 February 1967), 117–18.

Thompson, E. P. *The Making of the English Working Class.* Harmondsworth: Penguin Books, 1968.

Thompson, F. M. L. *English Landed Society in the Nineteenth Century.* London: Routledge and Kegan Paul, 1963.

Thompson, James A. *Count Rumford of Massachusetts.* New York: Farrar and Reinehart, 1935.

Thompson, S. P. *Michael Faraday.* London: Cassell and Company, 1898.

Thomson, H. Byerley. *The Choice of a Profession.* London: Chapman and Hall, 1857.

Thomson, Thomas. *History of the Royal Society, from its Institution to the end of the Eighteenth Century.* London: Robert Baldwin, 1812.

Thorpe, T. E. *Humphry Davy, poet and philosopher.* New York: Macmillan, 1896.

The Times, 2, 3, 4, 11, and 14 October 1844; 9 July 1855; 21 September 1931.

Townsend, Joseph. *A Dissertation on the Poor Laws.* London: C. Dilly in the Poultry, 1786. In J. R. McCulloch (ed.), *A Select Collection of Scarce and Valuable Economical Tracts, from the originals of Defoe, Elking, Franklin, Turgot, Anderson, Schomberg, Townsend, Burke, Bell, and Others.* London: Lord Overstone, 1859.

Treneer, Anne. *The Mercurial Chemist.* London: Methuen, 1963.

Tyndall, John. *Faraday as a Discoverer.* New York: Thomas Y. Crowell, 1961 reprint.

Underwood, E. A. (ed.). *A History of the Worshipful Society of Apothecaries 1617–1815.* London: Oxford University Press, 1963.

Valentine, Alan. *Lord George Germain.* Oxford: Clarendon Press, 1962.

Vernon, K. D. C. 'The Foundation and Early Years of the Royal Institution', *Proceedings of the Royal Institution of Great Britain*, XXXIX (1963), 364–402.

Viner, Jacob. 'Bentham and J. S. Mill: the Utilitarian Background', *The*

American Economic Review, XXXIX (1949), 360–82.

———— 'Intellectual History of Laissez-Faire', *The Journal of Law and Economics*, III (1960), 45–69.

Von Raumer, Frederick. *England in 1835*. 3 vols. London: John Murray, 1836.

Walsh, J. D. 'The Magdalene Evangelicals', *Church Quarterly Review*, CLIX (1958), 499–511.

Ward, J. R. *The Finance of Canal Building in Eighteenth-Century England*. London: Oxford University Press, 1974.

Warren, Samuel. *A Popular and Practical Introduction to Law Studies*. London: A. Maxwell, 1835.

Watson, J. S. *The Reign of George III*. Oxford: Clarendon Press, 1960.

Watson, James A. S., and Hobbs, May E. *Great Farmers*. London: Faber and Faber, 1951.

Watson, Robert S. *The History of the Literary and Philosophical Society of Newcastle-upon-Tyne (1793–1896)*. London: Walter Scott, 1897.

Webb, R. K. *Modern England*. New York: Dodd, Mead, 1968.

Webb, W. W. 'Peter Mark Roget', *DNB*, XLIX (1897), 149–51.

Weld, Charles R. *A History of the Royal Society*. 2 vols. London: John W. Parker, 1848.

Weston, W. J. *The County of Durham*. Cambridge University Press, 1914.

White, R. J. 'The Testament of Sir Humphry Davy', *History Today*, III (1953), 100–6.

———— *Waterloo to Peterloo*. Harmondsworth: Penguin Books, 1957.

Wiebe, R. H. *The Search for Order 1877–1920*. New York: Hill and Wang, 1968.

Wilberforce, Robert I. and Samuel. *The Life of William Wilberforce*. 5 vols. London: John Murray, 1838.

Williams, Gwyn. 'The Concept of "Egemonia" in the Thought of Antonio Gramsci: Some Notes on Interpretation', *The Journal of the History of Ideas*, XXI (1960), 586–99.

Williams, L. Pearce. 'Humphry Davy', *Scientific American*, CCII (June 1960), 106–16.

———— *Michael Faraday*. New York: Basic Books, 1965.

———— *The Origins of Field Theory*. New York: Random House, 1966.

———— (ed.). *Selected Correspondence of Michael Faraday*. 2 vols. Cambridge University Press, 1971.

Wilson, J. A. *The Chemistry of Leather Manufacture*. 2nd ed. 2 vols. New York: The Chemical Catalog Company, 1928–9.

Wood, Alexander, and Oldham, Frank. *Thomas Young*. Cambridge University Press, 1954.

Woodcock, George. *The British in the Far East*. London: Weidenfeld and Nicolson, 1969.

Wright, Charles, and Fayle, C. E. *A History of Lloyds*. London: Macmillan, 1928.

Wrigley, E. A. 'The Process of Modernization and the Industrial Revolution in England', *Journal of Interdisciplinary History*, III (1972), 225–60.

Young, Arthur. *The Autobiography of Arthur Young*. Ed. M. Betham-Edwards. London: Smith, Elder, 1898.
——— (ed.). *Annals of Agriculture, and other Useful Arts*. 46 vols. 1784–1815.
Young, G. M. *Victorian England: Portrait of an Age*. 2nd ed. New York: Oxford University Press, 1964.
Young, Thomas. *A Course of Lectures on Natural Philosophy and the Mechanical Arts*. 2 vols. London: J. Johnson, 1807.

Index

Page numbers followed by 'n' indicate that the reader should consult the footnote. Illustrations are shown by means of an asterisk.

The reader's attention is drawn to entries such as Faraday: these can occur under 'F' for Faraday, while books on his life would be under 'M' when dealing with Michael Faraday.

The letter-by-letter system of alphabetization has been adopted.